# Convex Optimization Techniques

# for
# Geometric Covering Problems

**Inaugural-Dissertation**

zur

Erlangung des Doktorgrades

der Mathematisch-Naturwissenschaftlichen Fakultät

der Universität zu Köln

vorgelegt von

## Jan Hendrik Rolfes

aus Neuwied

Köln, 2019

**Berichterstatter (Gutachter):**

| | |
|---|---|
| Prof. Dr. Frank Vallentin, | Universität zu Köln |
| Prof. Dr. David Gross, | Universität zu Köln |
| Prof. Dr. Cordian Riener, | University of Tromsø |

**Tag der mündlichen Prüfung:**  28.03.2019

© 2021, Jan Hendrik Rolfes
Herstellung und Verlag: BoD – Books on Demand, Norderstedt
ISBN: 9783754346754

# Preface

I enjoyed to spend my last four years on the present thesis and to work with great colleagues. First, I would like to take the opportunity to thank my advisor Frank Vallentin, who made this possible. He guided me through the process of writing this thesis and always had an open door to provide fruitful impulses. Thank you very much for everything.

I also want to thank Cordian Riener, who worked with me on the last chapter of this thesis and advised me during my stay in Tromsø. I enjoyed our discussions about mathematics and beyond. Thanks for your support and for a great time in Norway.

Next, I am very grateful that I was able to collaborate with Maria Dostert, Stefan Krupp, Frank Vallentin, Cordian Riener and David de Laat. The work with Frank, Cordian and David forms an important part of my thesis.

Another part of this thesis is formed by the Bachelor's thesis of Tuna Acisu. I enjoyed a lot co-supervising you and I hope you did the same.

I further want to thank my colleagues Fabricio Caluza Machado, Davi de Castro Silva, Maria Dostert, Anna Gundert, Frederik von Heymann, Annette Koenen, Stefan Krupp and Marc Zimmermann for creating a supportive environment that made me enjoying my way to the office. In particular, I want to thank Maria for teaching me how to make the best coffee in Cologne, Fred for giving me advice on two theses and Stefan for making me laugh. Additionally, I would like to thank Philippe for sharing his office with me.

A big thanks to Davi de Castro Silva, Maria Dostert, Anna Gundert, Frederik von Heymann, Kristina Hoffmann, Stefan Krupp, Julia Liesenklas and Lars Schmitz who were proof-reading parts of my thesis:

I owe a special thanks to Wolfgang Köhler, who enthused me with the "art of smart gaze" in grammar school. Thank you very much for encouraging me to study mathematics.

Last but not least, I want to thank my family and friends for supporting me in both academic and especially non-academic life. You make me feel happy.

i

# Kurzfassung

Die vorliegende Arbeit ist der Beginn einer Verallgemeinerung von Resultaten über spezifische Überdeckungen, wie etwa Überdeckungen des euklidischen Raumes oder Überdeckungen der Sphäre, zu einer Theorie auf kompakten metrischen Räumen. Insbesondere betrachtet man Überdeckungen eines kompakten metrischen Raumes $(X, d)$ durch Kugeln mit Radius $r$. Das Augenmerk liegt hierbei auf der minimalen Anzahl solcher Kugeln, welche benötigt wird um $X$ zu überdecken. Wir bezeichnen diese Anzahl mit $\mathcal{N}(X, r)$. Für endliche Räume $X$ entspricht dieses Problem einer Instanz des kombinatorischen SET COVER Problems, welches NP-vollständig ist. Wir beschreiben Approximationstechniken, basierend auf der Momentenmethode von Lasserre für endliche Graphen und verallgemeinern diese Techniken auf kompakte metrische Räume um untere und obere Schranken zu erhalten.

Die oberen Schranken in dieser Arbeit folgen aus der Anwendung eines Greedy-Algorithmus auf den Raum $X$. Die Approximationsgüte des Algorithmus erhalten wir durch eine Verallgemeinerung der Analyse von Chvátals Algorithmus für gewichtete SET COVER Probleme. Wir wenden den genannten Greedy-Algorithmus auf den sphärischen Fall $X = S^n$ an und erhalten die beste, nicht asymptotische Schranke von Böröczky und Wintsche. Weiterhin kann der Algorithmus genutzt werden, um Überdeckungen des euklidischen Raumes durch beliebige messbare Objekte mit nicht leerem Inneren zu bestimmen. Die Approximationsgüte dieser Überdeckungen stellt eine leichte Verbesserung der Schranken von Naszódi dar. Um untere Schranken zu erhalten entwickeln wir eine Folge von Schranken $\mathcal{N}^t(X, r)$, welche nach endlich vielen (bezeichnet mit $\alpha \in \mathbb{N}$) Schritten konvergiert:

$$\mathcal{N}^1(X, r) \leq \ldots \leq \mathcal{N}^\alpha(X, r) = \mathcal{N}(X, r).$$

Der Nachteil dieser Folge ist, dass die Schranken $\mathcal{N}^t(X, r)$ mit wachsendem $t$ immer schwieriger zu berechnen sind, da sie die Zielfunktionswerte unendlichdimensionaler konischer Programme sind, deren Anzahl an Bedingungen und Dimension der Kegel mit $t$ wachsen. Wir zeigen, dass diese Programme die Bedingung der starken Dualität erfüllen und leiten ein endlichdimensionales semidefinites Programm ab, welches darauf abzielt $\mathcal{N}^2(S^2, r)$ in beliebiger Präzision zu approximieren. Unsere Ergebnisse basieren teilweise auf der Momentenmethode von de Laat und Vallentin für das Packungsproblem auf topologischen Packungsgraphen. Jedoch müssen wir uns im Überdeckungsproblem um zwei Arten von

Bedingungen kümmern anstatt nur einer Art wie im Packungsproblem. Dies benötigt zusätzlichen Aufwand.

# Abstract

The present thesis is a commencement of a generalization of covering results in specific settings, such as the Euclidean space or the sphere, to arbitrary compact metric spaces. In particular we consider coverings of compact metric spaces $(X, d)$ by balls of radius $r$. We are interested in the minimum number of such balls needed to cover $X$, denoted by $N(X, r)$. For finite $X$ this problem coincides with an instance of the combinatorial SET COVER problem, which is NP-complete. We illustrate approximation techniques based on the moment method of Lasserre for finite graphs and generalize these techniques to compact metric spaces $X$ to obtain upper and lower bounds for $N(X, r)$.

The upper bounds in this thesis follow from the application of a greedy algorithm on the space $X$. Its approximation quality is obtained by a generalization of the analysis of Chvátal's algorithm for the weighted case of SET COVER. We apply this greedy algorithm to the spherical case $X = S^n$ and retrieve the best non-asymptotic bound of Böröczky and Wintsche. Additionally, the algorithm can be used to determine coverings of Euclidean space with arbitrary measurable objects having non-empty interior. The quality of these coverings slightly improves a bound of Naszódi.

For the lower bounds we develop a sequence of bounds $N^t(X, r)$ that converge after finitely (say $\alpha \in \mathbb{N}$) many steps:

$$N^1(X, r) \leq \ldots \leq N^\alpha(X, r) = N(X, r).$$

The drawback of this sequence is that the bounds $N^t(X, r)$ are increasingly difficult to compute, since they are the objective values of infinite-dimensional conic programs whose number of constraints and dimension of underlying cones grow accordingly to $t$. We show that these programs satisfy strong duality and derive a finite dimensional semidefinite program aiming to approximate $N^2(S^2, r)$ to arbitrary precision. Our results rely in part on the moment methods developed by de Laat and Vallentin for the packing problem on topological packing graphs. However, in the covering problem we have to deal with two types of constraints instead of one type as in packing problems and consequently additional work is required.

# Contents

CHAPTER ONE

# Introduction

## 1.1 A brief history of geometric coverings

Sometimes nature gives us a stunning intuition about elegant structures and arrangements. One famous example of such a structure, inspiring mathematicians working in extremal geometry, is the honeycomb arrangement. This arrangement provides two optimal configurations at once.

If we consider the incircle of each honeycomb, or hexagon, then this yields the densest possible arrangement of non-intersecting circles in the plane, essentially a result of the Norwegian mathematician Axel Thue [79] (see [31] for more details). The three-dimensional *sphere packing problem* became famous as the "Kepler conjecture" stated by Johannes Kepler in 1611. It was finally proved by Thomas C. Hales and Samuel P. Ferguson in 1996 in one of the first proofs of a major result for which computational optimization was used.

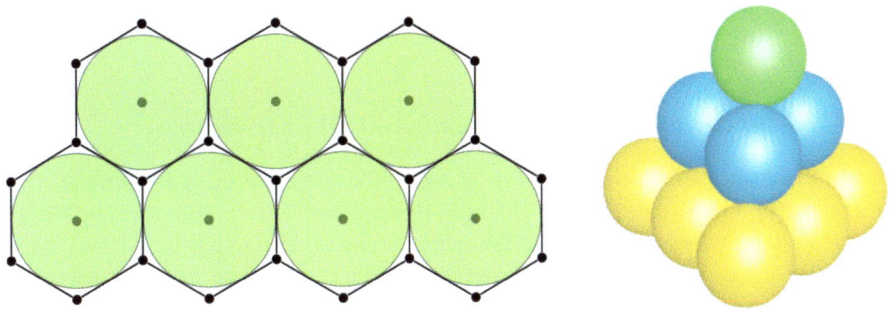

Figure 1.1: Two- and three-dimensional optimal packings.

For the purpose of this thesis, the most interesting arrangement stemming from honeycombs is provided by considering the circumcircles of each hexagon.

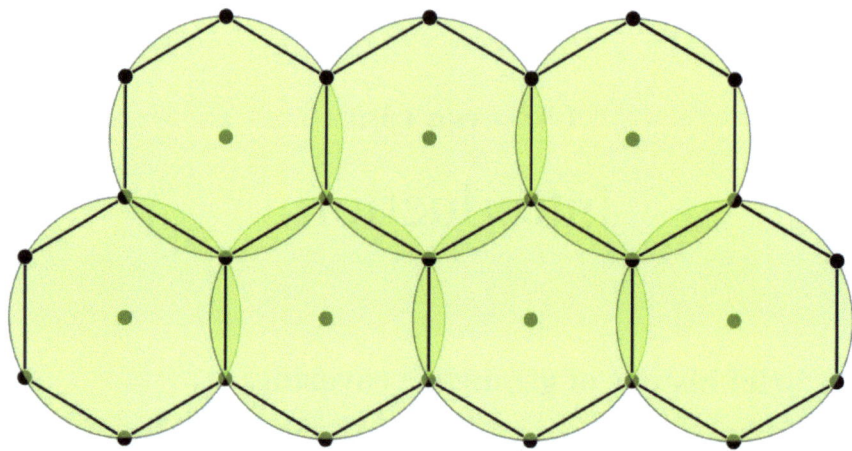

Figure 1.2: Thinnest covering of Euclidean space

These circumcircles cover the whole plane in a thinnest possible way, which was "well known" to Kershner before 1939 but he still quantified the statement and gave a proof in a classical paper [51]. The quantified statement reads as follows:

**Theorem 1.1.1.** *Let M denote a bounded set in the plane and let $N(\varepsilon)$ be the minimum number of circles of radius $\varepsilon$ which covers M. Then*

$$\lim_{\varepsilon \to 0} \pi \varepsilon^2 N(\varepsilon) = \frac{2\pi \sqrt{3}}{9} \lambda(\bar{M}),$$

*where $\bar{M}$ denotes the closure of M and $\lambda$ the standard Lebesgues measure.*

Here Kershner distinguished between $M$ and its closure $\bar{M}$ because this setting ensures the existence of a finite $N(\varepsilon)$ via the Heine-Borel theorem. The number $N(\varepsilon)$ in Theorem 1.1.1 can be seen as a first definition of the central object studied in this thesis, the covering number. We generalize the definition used by Kershner and define for a compact metric space $(X, d)$ and its corresponding closed balls $B(y, r) = \{x \in X : d(x, y) \leq r\}$ the *covering number* by:

$$N(X, r) = \inf_{Y \subseteq X} \left\{ |Y| : \bigcup_{y \in Y} B(y, r) = X \right\}.$$

Furthermore, if the space $(X, d)$ is equipped with a translation-invariant Borel measure $\mu$, such as the Lebesgues measure in Euclidean space, the measure $\mu(B(y, r))$ of a ball $B(y, r) \subseteq X$ does not depend on the location $y$ of the ball. The number

$$\Delta(X) := N(X, r) \cdot \frac{\mu(B(y, r))}{\mu(X)}$$

is called the *covering density* of $X$.

In the remainder of this thesis we will study approaches to approximate the covering number for a selection of different metric spaces. For the non-compact space $\mathbb{R}^n$, we consider a compact Euclidean ball $B(p, s) \subseteq \mathbb{R}^n$ and a covering $Y$ of Euclidean balls $B(y, r)$, i.e., $\bigcup_{y \in Y} B(y, r) = \mathbb{R}^n$. Let $Y' \subseteq Y$ denote the balls satisfying $B(y, r) \subseteq B(p, s)$. Then we say that $Y$ has a *lower covering density* of

$$\inf_{p \in \mathbb{R}^n} \liminf_{s \to \infty} \frac{\sum_{y \in Y'} \lambda(B(y, r))}{\lambda(B(p, s))}.$$

Furthermore, if for $Y$ and its corresponding subset $Y'$ the limit

$$\inf_{p \in \mathbb{R}^n} \lim_{s \to \infty} \frac{\sum_{y \in Y'} \lambda(B(y, r))}{\lambda(B(p, s))} = \liminf_{s \to \infty} \frac{\sum_{y \in Y'} \lambda(B(y, r))}{\lambda(B(p, s))}$$

exists, we call this limit the *covering density* of $Y$.

It is due to Groemer [38] that the above limit does not depend on the location $p$ of the ball and moreover for a minimal covering $Y$ we can replace $B(p, s)$ by any compact subset $sM \subseteq \mathbb{R}^n$ with positive measure. In such a case the *minimal covering density of $\mathbb{R}^n$*

$$\Delta(\mathbb{R}^n) := \inf_{Y \subseteq \mathbb{R}^n, \ Y \text{ covering}} \lim_{s \to \infty} \frac{\sum_{y \in Y'} \lambda(B(y, r))}{\lambda(sM)}$$

is attained at $Y$ and it is independent of $M$. Furthermore, since $Y'$ forms a covering of $(s - r)M$ for large $s$, we have $N((s - r)M, r) \leq |Y'|$ and consequently the limit reads

$$\Delta(\mathbb{R}^n) = \lim_{s \to \infty} \frac{N(sM, r) \cdot \lambda(B(y, r))}{\lambda(sM)},$$

which is again similar as in Kershner's theorem for the plane.

The research on coverings continued by considering compact domains in higher dimensional Euclidean spaces, but finding provably optimal arrangements turned out to be extremely difficult. In fact, even until now no arrangement is known to be optimal for Euclidean space of any dimension other than 2. In the following, the next results were achieved for a special class of covering arrangements, the arrangements formed by a *lattice* $L$. These arrangements are formed by a corresponding set $Y = L$, which is spanned by linearly independent vectors $v_1, \ldots, v_k$:

$$L = \left\{ \sum_{i=1}^{k} a_i v_i : a_i \in \mathbb{Z} \right\}.$$

The honeycomb arrangement is again a prime example, where the set $L$ describing the centers of each incircle and circumcircle is denoted by (see also Figure 1.2)

$$L = \left\{ a_1 \begin{pmatrix} 1 \\ 0 \end{pmatrix} + a_2 \begin{pmatrix} \frac{1}{2} \\ \frac{1}{2}\sqrt{3} \end{pmatrix} : a_i \in \mathbb{Z} \right\}.$$

By restricting to lattice coverings, i.e., lattices $L \subseteq \mathbb{R}^n$ with the property $\bigcup_{y \in L} B(y, r) = \mathbb{R}^n$, Bambah [7] considered the *lattice covering density*

$$\Delta_\Lambda(\mathbb{R}^n) := \min_{L \subseteq \mathbb{R}^n: \, L \text{ lattice covering}} \lim_{s \to \infty} \frac{\sum_{y \in L: \, B(y,r) \subseteq sM} \lambda(B(y,r))}{\lambda(sM)},$$

for any compact subset $M \subseteq \mathbb{R}^3$ and computed $\Delta_\Lambda(\mathbb{R}^3) = 1.464....$ The fact that the minimum among all the lattices is attained is a consequence of the Mahler selection theorem (see, e.g., [42]). Further optimality results for lattice arrangements in Euclidean space of dimension 4 (conjectured by Bambah, proven by Delone and Ryškov [25]) and dimension 5 (see Ryškov and Baranovskiĭ [72]) were obtained later. The lattices providing uniquely optimal configurations for dimensions $n = 2, \ldots, 5$ are the $A_n^*$ lattices, whereas the optimum for the case $n = 6$ is not attained at $A_6^*$. In general the lattice covering problem is still open for the cases $n \geq 6$. We refer to [15] for further details.

Due to the fact that $\Delta(\mathbb{R}^n) \leq \Delta_\Lambda(\mathbb{R}^n)$, Bambah computed a possibly sharp upper bound for the covering density of $\mathbb{R}^3$, and every thin (lattice) covering provides an upper bound on the minimal covering density of $\mathbb{R}^n$ accordingly. For the covering density of $\mathbb{R}^n$ general upper bounds have been proposed, as can be seen in Table 1.1.

| bound | author | method |
|-------|--------|--------|
| $n \ln n + n \ln \ln n + 5n + o(n)$ | Rogers [66] | probabilistic |
| $\left( \frac{1}{2} + o(1) \right) n \ln n$ | Dumer [29] | prob. & asymptotic |
| $n \ln n + n \ln \ln n + n + o(n)$ | Fejes-Tóth [30] | probabilistic |
| $n \ln n + n \ln \ln n + n + o(n)$ | Rolfes and Vallentin [70] | deterministic |

Table 1.1: Upper bounds for the covering density $\Delta(\mathbb{R}^n)$, additional terms in $o(n)$ are not stated due to simplicity

In 1959, Coxeter, Few and Rogers [16] improved the trivial *volume bound*:

$$\mathcal{N}(X, r) \geq \frac{\mu(X)}{\mu(B(y, r))} \tag{1.1}$$

and invented a bound based on simplices in the Euclidean space. Their bound holds in every dimension of Euclidean space, and in particular for $\mathbb{R}^3$ it yields a lower bound on the covering density of $1.431...$ , which is very close to Bambah's upper bound of $1.464...$ . Additionally, the bound in dimension 4 is also close to the upper bound given by the optimal lattice arrangement, and asymptotically we can state for the covering density of $\mathbb{R}^n$:

$$\lim_{r \to 0} \frac{N(M, r)\lambda(B(y, r))}{\lambda(M)} \geq \frac{n}{e\sqrt{e}},$$

whenever $n$ is sufficiently large.

Another direction of research in this area is to examine the covering number for compact metric spaces, which will be the central objective of this thesis. A good starting point to gain intuition is to investigate the upper and lower bounds for $N(S^n, r)$ on the unit sphere equipped with Euclidean standard metric. In the next section we give an overview of previous results. At first, we consider upper bounds, and later, previous lower bounds for $N(S^n, r)$.

## 1.2 Upper and lower bounds on the sphere

The spherical covering problem on $S^n$ can be considered as covering the compact metric space $(S^n, d)$, where the metric $d(x, y) := \arccos(x \cdot y)$ is the spherical distance. We equip this space with the unique rotational invariant probability measure $\omega$ and consider the problem

$$N(S^n, r) = \inf_{Y \subseteq S^n} \left\{ |Y| : \bigcup_{y \in Y} B(y, r) = S^n \right\}.$$

Four years after his work on lower bounds, Rogers [67] gave a first upper bound on $N(S^n, r)$ for dimensions $n \geq 8$ of

$$N(S^n, r)\omega(B(y, r)) \leq n \ln n + n \ln \ln n + n \ln \frac{1}{r} + o(n).$$

Starting in 2000 with the work of Böröczky and Wintsche [11], the upper bounds were further improved significantly. For large values of $n$ Dumer [29] provided new upper bounds. A more general approach by Naszódi [60] led to better bounds in lower dimensions. The current best bound for the non-asymptotic case is due to Rolfes and Vallentin [70] and will be content of Section 4.3.1. We give an overview of these improvements in Table 1.2 below.

| bound | author | method | restrictions |
|---|---|---|---|
| $n \ln n + n \ln \ln n + n \ln \frac{1}{r} + o(n)$ | Rogers [67] | probabilistic | $n \geq 8$ |
| $n \ln n + n \ln \ln n + 2n + o(n)$ | Böröczky and Wintsche [11] | probabilistic | none |
| $\left(\frac{1}{2} + o(1)\right) n \ln n$ | Dumer [29] | probabilistic | $n \to \infty$ |
| $n \ln n + n \ln \ln n + 2n + o(n)$ | Naszódi [60] | deterministic | none |
| $n \ln n + n \ln \ln n + n + o(n)$ | Rolfes and Vallentin [70] | deterministic | none |

Table 1.2: Upper bounds for the covering density $N(S^n, r)\omega(B(y, r))$, additional terms in $o(n)$ are not stated due to simplicity

The recent developments of Naszódi as well as of Rolfes and Vallentin rely on a greedy algorithm that iteratively chooses the balls covering the maximum measure of yet uncovered space. This greedy algorithm has been analyzed in the finite setting of the SET COVER problem, which is a fundamental problem in combinatorial optimization and content of Section 3.1.2.

The weighted SET COVER problem is defined as follows: Fix a number of elements $1, \ldots, m$. Given a collection $S_1, \ldots, S_n \subseteq \{1, \ldots, m\}$ and given costs $c_1, \ldots, c_n$, the task is to find a set of indices $I \subseteq \{1, \ldots, n\}$ such that $\bigcup_{i \in I} S_i = \{1, \ldots, m\}$ and $\sum_{i \in I} c_i$ is as small as possible. For determining upper bounds, Chvátal [13] showed that the greedy algorithm gives a $(\ln m + 1)$-approximation for the SET COVER problem; previously Johnson [47], Stein [78] and Lovász [57] proved similar results for the case of uniform costs $c_1 = \ldots = c_n = 1$.

On the one hand Naszódi [60] applied the results of Lovász [57] on SET COVER directly after choosing a finite $\varepsilon$-net. On the other hand Rolfes and Vallentin [70] transferred Chvátal's argument from the finite SET COVER setting to the infinite setting of compact metric spaces such as the sphere and got slightly better constants. This result is part of Chapter 4, the application to Naszódi's theory is illustrated in Section 4.3.3.

As mentioned in the last section, a trivial lower bound for the spherical covering problem is the volume bound, i.e.,

$$\frac{1}{\omega(B(y, r))} \leq \mathcal{N}(S^n, r).$$

In Example 6.3 of [11], Böröczky and Wintsche give an adjusted version of the Coxeter-Few-Rogers bound [16] for Euclidean space based on spherical simplices, stating that for sufficiently small radii $r$ we can bound the covering number from below by

$$\frac{c \cdot n}{\omega(B(y, r))} \leq \mathcal{N}(S^n, r),$$

where $c > 0$ is an absolute constant. These bounds are the best bounds known so far except for the special case of the two-dimensional sphere $S^2$. In this particular case Fejes-Tóth gave a lower bound in his classical book [31] based on spherical simplices. After a few elementary calculations, it states

$$\frac{2 \operatorname{arccot}\left( \sqrt{3}(1 - 2\omega_r)\right)}{\operatorname{arccot}\left( \sqrt{3}(1 - 2\omega_r)\right) - \frac{\pi}{6}} \leq \mathcal{N}(S^2, r).$$

On the other hand, Sloane published a number of "putatively optimal" coverings on his webpage [76] (see also Table 5.1), computed along with Hardin and Smith. We illustrate these upper and lower bounds for different angles $\varphi = \arccos r$ in the following plot:

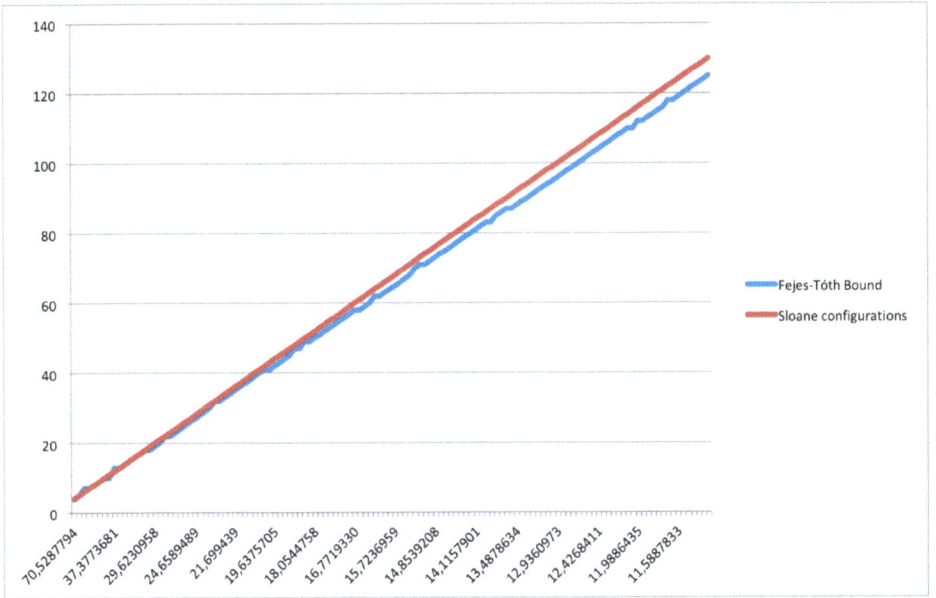

Figure 1.3: Comparison upper vs. lower bounds for $S^2$

For angles $< 11.588...$ degrees, Hardin, Sloane and Smith don't provide further configurations, due to the limits of computational power. To extend, generalize, and improve these bounds was the starting point for the present thesis.

The covering number occurs in several, sometimes unexpected fields. In the next section we give a handful of examples – without any claim to be exhaustive.

## 1.3   Related problems and applications

Although it originated from the special case of covering the Euclidean space as illustrated in Section 1.1, determining the covering number is a fundamental problem in metric geometry (see for example the classical book [68] of Rogers). After giving a brief overview of other covering problems, this section is devoted to direct applications of the covering number.

Besides the spherical and Euclidean cases, another well investigated problem is the covering number for the $n$-dimensional unit ball $\mathcal{N}(B(0,1),r)$. In his already mentioned work [67], Rogers also considered this problem and gave upper bounds for $\mathcal{N}(B(0,1),r)$. Verger-Gaugry improved these bounds for $r > \frac{\ln n}{n}$ and Dumer [29] provided asymptotic upper bounds.

For a compact and convex body $K \subseteq \mathbb{R}^n$, a similar covering number is part of active research in the context of the famous Levi-Hadwiger problem:

*Can a compact and convex body $K \subseteq \mathbb{R}^n$ be covered by at most $2^n$ homothetic smaller copies of itself?*

For $n = 2$ Levi [56] gave a positive answer. It is conjectured that the $n$-dimensional hypercube is the only convex body for which exactly $2^n$ smaller copies are needed. In recent papers, Artstein-Avidan and Raz [4] and Artstein-Avidan and Slomka [5] observed the strong relation between geometric coverings and SET COVER. They used the results of Lovász [57] to prove new results on geometric coverings, in particular to retrieve upper bounds for the Levi-Hadwiger problem: If $K$ is centrally symmetric, it can be covered by at most $2^n(n \ln n + n \ln \ln n + 5n)$ smaller copies of itself.

A remarkable fact is that the considered problems have one commonality: We always fix the size of the balls $B(y, r)$ to cover a fixed metric space $(X, d)$ and minimize the number $N$ of these balls needed to cover $X$. Another very active field of research asks for the following: If we fix a configuration of $N$ points $Y_N = \{y_1, \dots, y_N\} \subseteq X$, what is the minimal radius of the balls $B(y_i, r)$ needed to cover $X$, i.e.,

$$\rho(Y_N, X) = \min_{r \in \mathbb{R}} \left\{ r : \bigcup_{i=1}^N B(y_i, r) \supseteq X \right\} = \max_{x \in X} \min_{i \in [N]} d(y_i, x)?$$

The number $\rho(Y_N, X)$ is often referred to as *covering radius* or mesh norm. This is one of the few situations in which there are results known that hold for arbitrary compact metric spaces. In particular, Frostman (see [58], Theorem 8.17) showed for any compact metric space $(X, d)$ equipped with a finite $n$-dimensional Haussdorf measure, that there exists a positive constant $C$ such that for every $Y_N \subseteq X$ we obtain a lower bound of

$$\rho(Y_N, X) \geq \frac{C}{N^{\frac{1}{n}}}, \text{ if } N \geq 1.$$

However, the case drawing the most interest is when the points in $Y_N$ are distributed randomly on $X$. For $X = S^n$ equipped with the usual spherical distance and $Y_N$ distributed with respect to the surface measure, Reznikov and Saff [64] showed

$$\lim_{N \to \infty} \mathbb{E}\,\rho(Y_N, S^n) \left( \frac{N}{\ln N} \right)^{\frac{1}{n}} = \left( 2\sqrt{\pi} \frac{\Gamma\left(\frac{n+2}{2}\right)}{\Gamma\left(\frac{n+1}{2}\right)} \right)^{\frac{1}{n}}.$$

This result can be compared to our results in Chapter 4, where we fix the radius $\rho$ and approximate the corresponding covering number $N$, instead of fixing the number $N$ and approximate $\rho$. In particular, if one has $N = N(X, r)$ for every radius $r$, then one can derive the minimal covering radius $\min_{Y_N} \rho(Y_N, X)$ over all configurations with fixed size $N$ and vice versa.

The covering problem has many applications in various fields: compressive sensing [34], approximation theory and machine learning [17] — to name a few. Besides of these areas, there are applications in probability theory and quantum computing on which we elaborate in the next paragraphs.
In probability theory the logarithm of the covering number of metric spaces was defined by Kolmogorov in the 1960s as *metric entropy* and has several applications in this field as can be seen, e.g., in the book of M. Ledoux and M. Talagrand [55]. A very fruitful application arises from the analysis of the regularity properties of Gaussian processes with the help of the covering number. We follow [55] to elaborate on it.
We recall a couple of basic facts first. A Gaussian random variable $G'$ is a real valued random variable in $L^2(\Omega, \mathcal{A}, P)$, where $\Omega$ is called the *ground set*, $\mathcal{A} \subseteq \mathcal{P}(\Omega)$ is a sigma-algebra and $P$ is a probability measure on $\Omega$ stemming from a normal distribution. We will focus on *centered* Gaussian variables, where

$$\mathbb{E}(G') = \int_\Omega G'(\omega) dP(\omega) = 0.$$

We define a Gaussian process $G = (G_t)_{t \in T}$, indexed by a set $T$ as a collection of random variables $G_t$, $t \in T$, where every finite linear combination $\sum_i \alpha_i G_{t_i}$, $\alpha_i \in \mathbb{R}$, $t_i \in T$ is a centered Gaussian variable. Such a Gaussian process induces a pseudometric $d_G(s, t) :=\|G_t - G_s\|_2$ on $T$. The relation between such Gaussian processes and the covering number is on the one hand pointed out by the Sudakov minoration theorem:

**Theorem 1.3.1** (Sudakov). *Let $G = (G_t)_{t \in T}$ be a Gaussian process inducing a pseudometric $d_G$ on $T$. Then there exists a constant $K > 0$ such that*

$$r \left(\ln \mathcal{N}(T, r)\right)^{\frac{1}{2}} \leq K \sup\{ \mathbb{E} \sup_{t \in F} G_t : F \text{ is finite in } T \} \quad \text{for every } r > 0.$$

Since the results in this thesis apply to metric spaces in contrast to pseudometric spaces $(T, d_G)$, we recall the fact that the corresponding quotient space $T/\sim$, where $s \sim t$ if $d_G(s, t) = 0$, is a metric space equipped with metric $d(\pi(s), \pi(t)) = d_G(s, t)$ and $\pi : T \to T/\sim$ denotes the canonical quotient map. Additionally, the covering numbers $\mathcal{N}(T, r)$ and $\mathcal{N}(T/\sim, r)$ coincide, implying that any lower bound for $\mathcal{N}(T/\sim, r)$ provides a lower bound applicable to Sudakov's Theorem. On the other hand, there exists a theorem of Dudley to bound a Gaussian process $G$ from above:

**Theorem 1.3.2.** *Let $G = (G_t)_{t \in T}$ be a centered Gaussian process inducing a pseudometric $d_G$ on $T$. Then*

$$\mathbb{E} \sup_{t \in T} G_t \leq 24 \int_0^\infty \left(\ln \mathcal{N}(T, r)\right)^{\frac{1}{2}} dr.$$

It is worth noting that the results in the upcoming Chapter 4 on $N(X, r)$ for a compact metric space $X$ will provide explicit upper bounds for $\mathbb{E}\sup_{t \in T} G_t$.

In quantum computing a major question is how one can approximate an arbitrary single qubit gate to an accuracy $\varepsilon$ with a fixed finite set of quantum gates. The gates in this model are represented by unitary matrices $U$ lying in the compact metric space $(\mathrm{SU}(2), D)$, where $\mathrm{SU}(2)$ denotes the special unitary group of degree 2, a compact group consisting of all $2 \times 2$ unitary matrices with determinant 1. The *trace distance $D$* is defined by

$$D(U, V) := \mathrm{Tr}\left( \sqrt{(U - V)^H (U - V)} \right)$$

the trace of the positive square root of the product of $(U - V)$ and its adjoint matrix $(U - V)^H$. The following definition helps us to quantify the posed question:

For $\varepsilon > 0$ we define an *$\varepsilon$-net* of $\mathrm{SU}(2)$ as a subset $W \subseteq \mathrm{SU}(2)$ such that, for every matrix $U \in \mathrm{SU}(2)$ there is a matrix $V \in W$ such that $D(V, U) < \varepsilon$ holds.

The covering number $N(\mathrm{SU}(2), \varepsilon)$ thus provides the minimal amount of gates needed in an $\varepsilon$-net. In particular, $N(\mathrm{SU}(2), \varepsilon)$ provides a lower bound on the size of the set $\mathcal{G}_l$ used in the following *Solovay-Kitaev theorem*.

**Theorem 1.3.3** (Solovay-Kitaev, [61]). *Let $\mathcal{G} \subseteq \mathrm{SU}(2)$ be a finite set of gates and*

$$\langle \mathcal{G} \rangle = \{ U \in SU(2) : \ \exists \, m \in \mathbb{N}, G_1, \ldots, G_m \in \mathcal{G} : \ U = G_1 \ldots G_m \}$$

*is dense in $SU(2)$, i.e., for every $U \in SU(2)$ and $\delta > 0$ there exists a $V \in \langle \mathcal{G} \rangle$: $D(U, V) < \delta$. Then, if we fix $\varepsilon > 0$,*

$$\mathcal{G}_l = \{ U \in SU(2) : \ \exists \, k < l, G_1, \ldots, G_k \in \mathcal{G} : \ U = G_1 \cdots \cdots G_k \}$$

*forms an $\varepsilon$-net of $SU(2)$ with $l = O\left( \log^c \left( \frac{1}{\varepsilon} \right) \right)$, where $c$ is a small constant, approximately equal to 2.*

We observe that $N(\mathrm{SU}(2), \varepsilon) \leq |\mathcal{G}_l|$. For more details on the Solovay-Kitaev theorem we refer to the book [61] of Nielsen and Chuang.

## 1.4   Outline of the thesis

In this thesis we approach the covering problem as follows: First, we construct coverings of $X$ to obtain upper bounds for the covering number $N(X, r)$, and second, we obtain lower bounds using semidefinite programming techniques. To guide the reader towards these results, the thesis contains five chapters, including the present introductory chapter. In Chapter 2 the basic mathematical techniques are explained which are used later in the following Chapters 3 to 5. In Chapter 3 we demonstrate both approaches to the covering problem, the construction of coverings and the semidefinite programming techniques, by considering the classical combinatorial problem SET COVER. The construction part is already published in [70] and will be described in Chapter 4, whereas the lower bounds are studied in Chapter 5.

## 1.4.1 Chapter 2: Techniques

For the topics treated in Chapters 4 and especially 5 we provide an introduction to convex optimization on general topological vector spaces. In particular, we focus on the duality of signed Radon measures and continuous functions acting on a compact metric space $X$ stemming from the Riesz representation theorem. Additionally, we provide some background on semidefinite programming and study under which conditions a semidefinite optimization program is computable in polynomial time.

In Section 2.2 we investigate quotient spaces on $X$ with respect to the group of isometries $Iso(X)$ acting on $X$ and give examples, pointing out the relation to the geometric problems we want to tackle. We further examine the topological properties of certain metric spaces $X$ and their corresponding function spaces $C(X)$ in order to apply the theorem of Arzelà-Ascoli to $Iso(X)$. This enables us to extend the results in the upcoming Chapter 4 to a wider field of applications.

In the last section we introduce the technique of symmetry reduction of semidefinite programs and sketch how the theorems of Peter-Weyl and Bochner apply to the geometric setting. Symmetry reduction plays a major role to finally achieve computable lower bounds for specific covering numbers.

## 1.4.2 Chapter 3: The Lasserre hierarchy on SET COVER

In Chapter 3 we study the SET COVER problem to illustrate the methods used in the remainder of this thesis in the setting of simple graphs. The chapter is based on lecture notes by Rothvoß [71], elaborated in a Bachelor's thesis of Acisu [1], that demonstrate the power of Lasserre's [54] semidefinite programming hierarchy.

We recall that the classical combinatorial problem SET COVER addresses the task of covering a set of elements $\{1, \ldots, m\}$ with certain fixed subsets $S_1, \ldots, S_n$ at minimal cost. To be more precise: One picks subsets $S_i$, indexed by $I \subseteq \{1, \ldots, n\}$, with $\bigcup_{i \in I} S_i = \{1, \ldots, m\}$ and $\sum_{i \in I} c_i$ is minimal. We can also formulate SET COVER as a 0/1-integer problem:

$$SC = \min \sum_{i=1}^{n} c_i x_i : \tag{1.2}$$

$$\sum_{i:\, k \in S_i} x_i \geq 1 \quad \text{for every } k \in [m],$$

$$x_i \in \{0, 1\} \quad \text{for every } i \in [n].$$

The integer program asks whether $S_i$ should be included in our covering ($x_i = 1$) or not ($x_i = 0$) such that every $k \in [m]$ is contained in at least one included set $S_i$ with $x_i = 1$ and such that the costs are minimal. The SET COVER problem was among the first problems for which NP-completeness was shown in 1972 [49], making it one of Karp's famous 21 problems. Dinur and Steurer [27] even showed that for every $\varepsilon > 0$ it is NP-hard to find an approximation to the SET COVER problem within a factor of $(1 - \varepsilon)\ln m$.

On the positive side we have already stated in Section 1.2 that a simple greedy algorithm results in a solid upper bound. We will give a slight variation of Chvátal's [13] proof for the following theorem to point out the strong relation between linear optimization and the greedy algorithm.

**Theorem 1.4.1.** *Let c be the cost of a cover J returned by the greedy algorithm. Then*

$$c \leq H(M) \cdot SC$$

*with $H(M)$ being the harmonic series $H(M) = \sum_{j=1}^{M} \frac{1}{j} \leq \ln M + 1$ and $M = \max_{i \in \{1,...,n\}} |S_i|$ being the size of the largest subset $S_i$.*

Note that together with the results of Dinur and Steurer, Theorem 1.4.1 shows that the greedy algorithm is optimal if P $\neq$ NP. In the remainder of Chapter 3 we focus on the Lasserre hierarchy to compute lower bounds for SET COVER and illustrate how the corresponding semidefinite relaxations have been used by Chlamtác, Friggstad and Georgiou [12] to improve the upper bound given by Theorem 1.4.1. For this reason we mainly follow Rothvoß' lecture notes [71] to give a full proof for

**Theorem 1.4.2.** *For a fixed $\varepsilon$ with $0 < \varepsilon < 1$, one can find an approximation for SET COVER with cost at most $((1 - \varepsilon)\ln m + o(1)) \cdot SC$ in time $n^{O(m^\varepsilon)}$.*

The relation of SET COVER to geometric covering problems is rather direct, as a possible approach to achieve further bounds for $N(S^n, r)$ would be to sample the underlying geometric structure and work with the corresponding SET COVER problem. However, the results presented in this thesis do not apply the approaches for SET COVER directly, but instead try to generalize the methods used by Chvátal and Lasserre to geometric covering problems. The intention behind this approach is to discretize only at the very end to be able to exploit the problem structure. Hence, it is the main aim of this thesis to find analogues of Theorems 1.4.1 and 1.4.2 in the infinite setting.

## 1.4.3 Chapter 4: Covering compact metric spaces greedily

We consider the problem of finding upper bounds for the covering number of a compact metric space $(X, d)$ equipped with a probability measure $\omega$; a Borel measure normalized by $\omega(X) = 1$. We will assume that this probability measure behaves homogeneously on balls and is non-degenerate, i.e., it satisfies the following two conditions:

(a)  $\omega(B(x, s)) = \omega(B(y, s))$ for all $x, y \in X$, and for all $s \geq 0$,

(b)  $\omega(B(x, \varepsilon)) > 0$ for all $x \in X$, and for all $\varepsilon > 0$.

By (a) the measure of a ball only depends on the radius $s$ and not on the center $x$, so we simply denote $\omega(B(x, s))$ by $\omega_s$. The main result of [70] is as follows.

**Theorem 1.4.3.** *Let $(X, d)$ be a compact metric space with probability measure $\omega$. Then for every $\varepsilon$ with $r/2 > \varepsilon > 0$ the covering number satisfies*

$$\frac{1}{\omega_r} \leq N(X, r) \leq \frac{1}{\omega_{r-\varepsilon}} \left( \ln\left( \frac{\omega_{r-\varepsilon}}{\omega_\varepsilon} \right) + 1 \right).$$

Our proof is based on a greedy approach to covering. We iteratively choose balls which cover the maximum measure of yet uncovered space and prove that its running time does not exceed $\frac{1}{\omega_{r-\varepsilon}} \left( \ln\left( \frac{\omega_{r-\varepsilon}}{\omega_\varepsilon} \right) + 1 \right)$. An illustration how the greedy algorithm runs on the sphere $S^2$ is given in Figure 1.4:

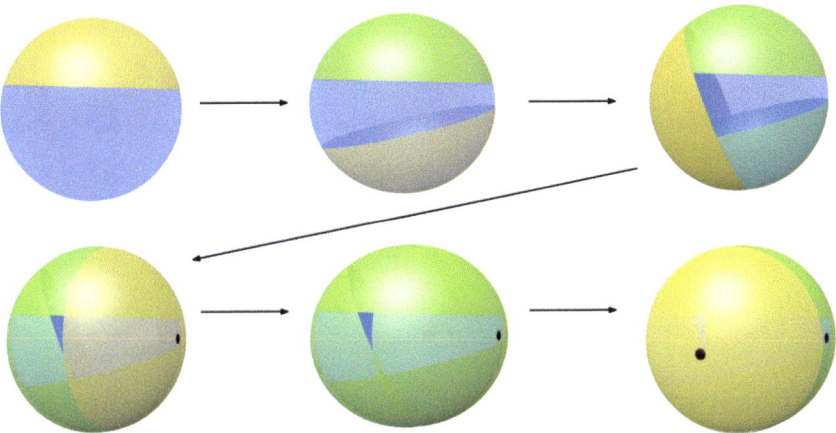

Figure 1.4: Illustration of the greedy algorithm for $N(S^2, 0.9)$

We further apply Theorem 1.4.3 to three concrete geometric situations and retrieve some of the best known asymptotic results, unifying many results on sphere coverings. In this context, by adding some further analysis, Theorem 1.4.3 yields

**Corollary 1.4.4.** *The covering density $N(S^n, r) \cdot \frac{\omega_r}{\omega(S^n)}$ of the $n$-dimensional sphere by spherical balls is at most*

$$\left( 1 + \frac{1}{v - 1} \right) (n \ln vn + 1) \quad \text{for all } v > 1.$$

*In particular, for $v = \ln n$, the covering density is at most*

$$n \ln n + n \ln \ln n + n + o(n).$$

Since our setup also includes the compact torus $X = \mathbb{R}^n / \mathbb{Z}^n$ with which we can tessellate the non-compact Euclidean space $\mathbb{R}^n$, we obtain coverings for $\mathbb{R}^n$ with the same density. As an example we reprove a bound of Fejes-Tóth [30].

**Corollary 1.4.5.** *The covering density of the n-dimensional Euclidean space by congruent balls is at most*

$$\left(1 + \frac{1}{v-1}\right)(n \ln vn + 1) \quad \text{for all } v > 1.$$

*In particular, for $v = \ln n$, the covering density is at most*

$$n \ln n + n \ln \ln n + n + o(n).$$

Moreover, by adding some further analysis, we were also able to slightly improve a bound of Naszódi [60]:

**Corollary 1.4.6.** *Let $K \subseteq \mathbb{R}^n$ be a bounded measurable set. Then there is a covering of $\mathbb{R}^n$ by translated copies of K of density at most*

$$\inf \left\{ \frac{\omega(K)}{\omega(K_{-\delta/2})} \left( \ln \left( \frac{\omega(K_{-\delta/2})}{\omega(B(0, \delta/2))} \right) + 1 \right) : \delta > 0, K_{-\delta} \neq \emptyset \right\},$$

*where $K_{-\delta} = \{x \in K : B(x, \delta) \subseteq K\}$ is the $\delta$-inner parallel body of K.*

### 1.4.4   Chapter 5: A Lasserre-type hierarchy for covering problems

The focus of Chapter 5 is to design a method to determine lower bounds for $N(X, r)$, that keeps the geometric structure of the compact metric space $X$ instead of working with a sample of $X$ and its corresponding SET COVER problem. We ultimately want to compute concrete lower bounds for the spherical covering number $N(S^n, r)$.

Recently, de Laat and Vallentin [23] used the following approach to successfully tackle the related sphere packing problem: They defined a *topological packing graph* on a Hausdorff topological space as a graph for which each finite clique is contained in an open clique, i.e., in an open subset of the vertex set where every two vertices are adjacent. For these graphs they developed a generalization of the Lasserre hierarchy for the corresponding INDEPENDENT SET problem, another prominent combinatorial optimization problem for finite graphs.

Our approach is similar to theirs, but since a covering, in contrast to a packing, does not automatically satisfy the Hausdorff property, we need further analysis and therefore can only treat the case of metric spaces or *distance graphs* instead of topological packing graphs. It is not known whether our approach can be generalized to topological spaces. In the following we aim to restate the covering number as a conic optimization problem using Dirac measures, define a generalized Lasserre hierarchy for this problem and give an overview of our results. Like in Section 1.1 we equip $X$ with a metric $d$ and thus define the corresponding covering number by

Figure 1.5: Spherical caps on $S^2$

$$N(X,r) = \min\left\{|Y| : Y \subseteq X, \bigcup_{y \in Y} B(y,r) = X\right\}. \tag{1.3}$$

For $X = S^n$ the balls $B(y,\varphi) = \{x \in S^n : x \cdot y \geq \cos\varphi\}$ are often referred to as *spherical caps* (see Figure 1.5).

In contrast to the integer program (5.3) it is not clear how to define a suitable objective function on the space of functions $V \to \mathbb{R}$ or $\mathbb{R}^V$ whenever $V$ is uncountable. A fact already observed by de Laat and Vallentin (see [23]). Instead we consider measures $\mu$ to define a program similar to (5.3). In particular we can reformulate $N(X,r)$ as

$$N(X,r) = \min\left\{\mu(X) : \mu = \sum_{y \in Y} \delta_y, \ \mu(B(\gamma x,r)) \geq 1 \text{ for all } \gamma \in \Gamma\right\}, \tag{1.4}$$

where $\Gamma \subseteq \text{Iso}(X)$ denotes a transitive group action and $\delta_y$ is the Dirac measure defined by

$$\delta_y(A) = \begin{cases} 1 & \text{if } y \in A, \\ 0 & \text{otherwise.} \end{cases}$$

In particular the finite sums of Dirac measures have the property that they lie in the cone $\mathcal{M}(\mathcal{B}(X))_{\geq 0}$ of nonnegative Radon measures acting on the Borel measures $\mathcal{B}(X)$.

Since the cone $\mathcal{M}(\mathcal{B}(X))_{\geq 0}$ is structurally too difficult, we aim for a simpler cone that inner-approximates $\mathcal{M}(\mathcal{B}(X))_{\geq 0}$ and still contains an optimal solution $\mu = \sum_{y \in Y} \delta_y$ for

(1.4). Over the course of Chapter 5 we show that for a fixed $\varepsilon > 0$ and $t \in \mathbb{N}$ the set $I_t := \{Y \subseteq X : |Y| \leq t, \ d(y, y') \geq \varepsilon \text{ for every } y, y' \in Y\}$ forms a compact metric space equipped with an extended Hausdorff metric. Moreover, we can apply a duality theory on the cone $\mathcal{M}(I_t)_{\geq 0}$, which we define below.

The set $C(I_t)$ of continuous real-valued functions on $I_t$ and the set $\mathcal{M}(I_t)$ of *signed Radon measures* form a topological dual pairing, where the topology is defined by the supremum norm on $C(I_t)$. This is a consequence of the Riesz representation theorem (see Chapter 2 or [8]). In particular, these dual spaces contain the cones $C(I_t)_{\geq 0}$ of nonnegative functions in $C(I_t)$ and its dual cone

$$\mathcal{M}(I_t)_{\geq 0} := \left\{ \mu \in \mathcal{M}(I_t) : \int_{I_t} f d\mu \geq 0 \text{ for all } f \in C(I_t)_{\geq 0} \right\}.$$

Similarly we can denote the space of *symmetric kernels* $C(I_t \times I_t)$ as the set of continuous functions $K : I_t \times I_t \to \mathbb{R}$, where $K(J, J') = K(J', J)$ for every $J, J' \in I_t$. We will define the cone of positive semidefinite kernels $C(I_t \times I_t)_{\geq 0}$ and denote its dual cone by $\mathcal{M}(I_t \times I_t)_{\geq 0}$ contained in the space of symmetric Radon measures $\mathcal{M}(I_t \times I_t)$, where $\mu(E, E') = \mu(E', E)$ holds for every pair of Borel sets $E, E'$. Let $I_{=1}$ denote the set of $J \subseteq I_t$ with $|J| = 1$, then our continuous counterpart for the Lasserre relaxation, which relaxes $\mathcal{N}(X, r)$ is defined as follows:

$$\mathcal{N}^t(X, r) = \inf \mu(I_{=1}) \tag{1.5}$$
$$\mu \in \mathcal{M}(I_{2t}/\Gamma)_{\geq 0},$$
$$\mu(\{\emptyset\}) = 1,$$
$$\tilde{A}_t \mu \in \mathcal{M}(I_t \times I_t)_{\geq 0},$$
$$\tilde{A}_t^{id} \mu \in \mathcal{M}(I_{t-1} \times I_{t-1})_{\geq 0},$$

where $\tilde{A}_t$ and $\tilde{A}_t^{id}$ are defined pointwise by their adjoint operators $\tilde{B}_t : C(I_t \times I_t) \to C(I_{2t}/\Gamma)$ and $\tilde{B}_t^{id} : C(I_{t-1} \times I_{t-1}) \to C(I_{2t}/\Gamma)$ defined by

$$\tilde{B}_t K(\pi(S)) := \int_\Gamma \sum_{\substack{J, J' \in I_t: \ J \cup J' = \gamma S}} K(J, J') d\lambda(\gamma)$$

$$\tilde{B}_t^{id} K(\pi(S)) := \int_\Gamma \sum_{\substack{x \in B_r, \ J, J' \in I_{t-1}: \ J \cup J' \cup \{x\} = \gamma S}} K(J, J') - \sum_{\substack{J, J' \in I_{t-1}: \ J \cup J' = \gamma S}} K(J, J') d\lambda(\gamma),$$

where $\lambda$ is the normalized Haar measure on $\Gamma$, $\pi : I_t \to I_t/\Gamma$ the quotient map and $B_r = B(e, r)$ is a ball centered at a fixed point $e \in X$. For $t = 1$ the summation over all $x, J, J'$ satisfying $J \cup J' \cup \{x\} = \gamma S$ boils down to the question whether $\gamma S \in I_{=1}$ and contained in $B_r$ or not. As we will show in Chapter 5, the programs above lead to the following non-decreasing sequence of bounds:

$$\mathcal{N}^1(X, r) \leq \mathcal{N}^2(X, r) \leq \ldots \leq \mathcal{N}^t(X, r) \leq \mathcal{N}(X, r) \tag{1.6}$$

for any value of $t$. Our two main results of Chapter 5 are that our continuous Lasserre hierarchy satisfies strong duality in each step and that the hierarchy converges after at most finitely many steps. First, the following theorem states the dual programming hierarchy and proves strong duality in each step of the hierarchy.

**Theorem 1.4.7.**

$$\mathcal{N}^t(X, r) = \sup y$$

$$y \in \mathbb{R}, \ K \in C(I_t \times I_t)_{\geq 0}, \ K' \in C(I_{t-1} \times I_{t-1})_{\geq 0}$$

$$\mathbb{1}_{I_{=1}/\Gamma}(Q) - y \mathbb{1}_{\{\pi(\emptyset)\}}(Q)$$

$$- \tilde{B}_t K(Q) - \tilde{B}_t^{id} K'(Q) \geq 0 \text{ for all } Q \in I_{2t}/\Gamma$$

After setting up a suitable hierarchy for the covering number, the second main result of Chapter 5 concerns the question whether the hierarchy *converges*, i.e., whether there is a finite step $T$ for which $\mathcal{N}^T(S^n, \varphi) = \mathcal{N}(S^n, \varphi)$ holds. It turns out that the *packing number*

$$\alpha := \alpha(X, \varepsilon) := \max \{|Y| : \ Y \subseteq X, \ B(y, \varepsilon) \cap B(y', \varepsilon) = \emptyset \text{ for every } y, y' \in Y\},$$

marks one such critical step in the hierarchy as we prove the following theorem.

**Theorem 1.4.8.** *Suppose $\mu$ is feasible for $\mathcal{N}^\alpha(X, r)$. Then the objective value $\mu(I_{=1})$ bounds the covering number $\mathcal{N}(X, r)$ from above.*

Together with the inequalities (1.6) this result proves the convergence of the given hierarchy after at most $\alpha$ steps. Combining the latter result with the fact that the first step coincides with the volume bound we achieve

$$\frac{\lambda(\Gamma)}{\lambda(\{\gamma \in \Gamma : \ \gamma\{y\} \in B_r\})} = \frac{\omega(X)}{\omega(B(x, r))} \leq \mathcal{N}^1(X, r) \leq \mathcal{N}^2(X, r) \leq \ldots \leq \mathcal{N}^\alpha(X, r) = \mathcal{N}(X, r),$$

where the dual programs provide possibly sharp lower bounds in each step.
Numerical bounds are still under investigation. However, for the two-dimensional spherical covering number $\mathcal{N}(S^2, r)$ we provide a finite dimensional semidefinite optimization program that aims to bound the number $\mathcal{N}^2(S^2, r)$ from below.

# CHAPTER TWO

# Techniques

The present Chapter 2 contains background material for the upcoming chapters. Experts in the field may skip this chapter or just read the parts that seem interesting to them.

## 2.1   Convex Optimization

The theory of convex optimization mainly started by Kantorovich's fundamental work on linear optimization in 1939 (see [48]). We quote Dantzig [19] to comment on the importance of his work:

*"Kantorovich should be credited with being the first to recognize that certain important broad classes of production problems had well-defined mathematical structures which, he believed, were amenable to practical numerical evaluation and could be numerically solved."*

It is important to note that in his work "Problem C" is essentially the first formulation of a linear program as we know it today (see [74]. Dantzig himself confirmed Kantorovich's believe that linear programs could be numerically solved by introducing the simplex algorithm in the summer of 1947, which was published in [18]. In October of the same year Dantzig and von Neumann met for the first time and von Neumann [81] was able to "immediately translate basic theorems in game theory into their equivalent statements for systems of linear inequalities"[1].

In linear optimization one optimizes over the cone $\mathbb{R}_{\geq 0}^n$ of nonnegative Euclidean vectors but it was discovered over the years that the main results of Kantorovich's theory hold in a more general setting. In this chapter we give a short overview of the theory of conic optimization, mainly following Barvinok's book [8].

---

[1][19]

We begin by considering topological vector spaces $E$ and $F$ and call a non-degenerate bilinear form $\langle \cdot, \cdot \rangle : E \times F \to \mathbb{R}$ a *duality* of $E$ and $F$ if the following two properties are satisfied:

- for every $f \in F$ the linear functional $\phi : E \to \mathbb{R}$ defined by $\phi = \langle \cdot, f \rangle$ is continuous and every continuous linear functional $\phi : E \to \mathbb{R}$ can be written as $\phi = \langle \cdot, f \rangle$ for some unique $f$,

- for every $e \in E$ the linear functional $\phi : F \to \mathbb{R}$ defined by $\phi = \langle e, \cdot \rangle$ is continuous and every continuous linear functional $\phi : F \to \mathbb{R}$ can be written as $\phi = \langle e, \cdot \rangle$ for some unique $e$.

The concept of having a duality of two topological vector spaces is a key ingredient in convex optimization. In particular if one optimizes over a convex cone $\mathcal{K}$, e.g., over the cone of nonnegative vectors $\mathcal{K} = \mathbb{R}^n_{\geq 0}$ as in linear optimization, the *dual cone* is often of great interest.

**Definition 2.1.1** (Dual cone). *Let $\langle \cdot, \cdot \rangle : E \times F \to \mathbb{R}$ be a duality and $\mathcal{K} \subseteq E$ be a cone. The dual cone $\mathcal{K}^*$ of $\mathcal{K}$ is*

$$\mathcal{K}^* := \{ f \in F : \langle e, f \rangle \geq 0 \text{ for every } e \in \mathcal{K} \}.$$

We call a cone $\mathcal{K}$ *self-dual* if $\mathcal{K}^* = \mathcal{K}$. Let us now consider two dualities $\langle \cdot, \cdot \rangle_1$ of topological vector spaces $E_1$ and $F_1$ and $\langle \cdot, \cdot \rangle_2$ of topological vector spaces $E_2$ and $F_2$. Additionally, we consider a linear operator $A : E_1 \to E_2$ and its *adjoint operator* $B : F_2 \to F_1$, that is uniquely defined as the linear operator for which

$$\langle A(x), y \rangle_2 = \langle x, B(y) \rangle_1$$

holds for every $x \in E_1$ and $y \in F_2$. For convex cones $\mathcal{K}_1 \subseteq E_1$ and $\mathcal{K}_2 \subseteq E_2$ and two fixed elements $c \in F_1$ and $b \in E_2$ we can define a *primal conic program* by

$$p^* = \inf_{x \in \mathcal{K}_1} \{ \langle x, c \rangle_1 : A(x) - b \in \mathcal{K}_2 \} \tag{2.1}$$

and its *dual conic program* by

$$d^* = \sup_{y \in \mathcal{K}_2^*} \{ \langle b, y \rangle_2 : c - B(y) \in \mathcal{K}_1^* \}. \tag{2.2}$$

It is important to note that a wide class of problems can be described as conic programs in the form (2.1) or (2.2). As we stated before, linear programming falls into this setup by setting $\mathcal{K}_1 = \mathcal{K}_2 = \mathbb{R}^n_{\geq 0}$ but also optimization over other convex cones is well-established. Of particular interest in the remainder of this thesis are the two pairs of convex cones:

- The cone $S_{\geq 0}^n$ of positive semidefinite matrices which is self-dual, meaning $(S_{\geq 0}^n)^* = S_{\geq 0}^n$.

- The cone of nonnegative Radon measures $M(X)_{\geq 0}$ on a compact metric space $X$ and its dual cone $C(X)_{\geq 0}$ of nonnegative continuous functions on $X$.

The strength of the theory above lies in the strong relation between primal and dual programs, which makes the following theorem arguably one of the most important statements in this whole thesis.

**Theorem 2.1.2** (Weak and strong duality). *Suppose we are given a pair of primal and dual conic programs. Let $p^*$ be the supremum of the primal and $d^*$ the infimum of the dual program.*

1. **Weak duality:** *Suppose $x$ is a feasible solution for the primal and $y$ is feasible for the dual conic program. Then,*
$$\langle x, c \rangle_1 \geq \langle b, y \rangle_2.$$
*In particular $p^* \geq d^*$.*

2. **Complementary Slackness:** *Suppose that the primal conic program attains its supremum at $x$, that the dual conic program attains its infimum at $y$, and that $p^* = d^*$. Then*
$$\langle x, c - B(y) \rangle_1 = 0.$$

3. **Optimality Criterion:** *Suppose that $x$ is a feasible solution of the primal conic program, $y$ is a feasible solution of the dual conic program and equality*
$$\langle x, c - B(y) \rangle_1 = 0$$
*holds. Then the supremum of the primal conic program is attained at $x$ and the infimum of the dual conic program is attained at $y$.*

4. **Strong duality:** *If the dual program is bounded from above and if $c - B(y) \in int(\mathcal{K}_1^*)$, $y \in int(\mathcal{K}_2^*)$ and the dual program attains its supremum, then $p^* = d^*$. If the primal program is bounded from below and if $x \in int(\mathcal{K}_1)$, $A(x) - b \in int(\mathcal{K}_2)$ and the primal program attains its infimum, then $p^* = d^*$.*

Condition 4 is known as *Slater's condition*. Another sufficient condition for strong duality will be helpful in proving Theorem 5.4.4. We further denote the difference $p^* - d^*$ as the *duality gap* of the pair of primal and dual programs.
As a toy example to illustrate the power of the above Theorem 2.1.2, we consider the cones $\mathcal{K}_1 = S_{\geq 0}^n$, $\mathcal{K}_2 = \{0\}$ and their dual cones $\mathcal{K}_1^* = S_{\geq 0}^n$, $\mathcal{K}_2^* = \mathbb{R}$ with respect to the Frobenius inner product $\langle A, B \rangle = \text{Tr}(A^T B)$ and the standard multiplication on $\mathbb{R}$. This allows us to apply the corresponding conic programs to the question of finding the largest or smallest eigenvalues of a symmetric matrix.

**Example 2.1.3** (Eigenvalue optimization). *The largest and smallest Eigenvalues of a symmetric matrix $C \in S^n$ can be expressed with the following semidefinite programs:*

$$\lambda_{max}(C) = \max_{X \in S^n_{\geq 0}} \{\langle C, X \rangle : Tr(X) = 1\} = \min_{y \in \mathbb{R}} \{y : yI_n - C \in S^n_{\geq 0}\},$$

$$\lambda_{min}(C) = \min_{X \in S^n_{\geq 0}} \{\langle C, X \rangle : Tr(X) = 1\} = \max_{y \in \mathbb{R}} \{y : C - yI_n \in S^n_{\geq 0}\}.$$

*Here strong duality holds since Slater's condition holds on the right hand sides.*

An immediate observation of the formulation as an optimization problem is the fact that due to weak duality, every feasible solution $y \in \mathbb{R}$ provides an upper bound for the maximal Eigenvalue, respectively a lower bound for the minimal Eigenvalue. Due to Theorem 2.1.2 this pattern holds in the general setup, i.e., we can use the solutions to the dual program to determine lower bounds for the primal and vice versa, which will be a key technique thoughout this thesis.

Furthermore, if one considers a program

$$p = \inf_{x \in \mathcal{K}_1} \{\langle c, x \rangle_1 : x \in S\}$$

with (possibly non-conic) constraints $x \in S$, then we call a program

$$p' = \inf_{x \in \mathcal{K}_1} \{\langle c, x \rangle_1 : x \in S'\}$$

with $S' \supseteq S$ a *relaxation* of $p$ and $p' \leq p$ provides a lower bound on $p$. If the remaining constraints on $x \in S'$ are conic, i.e., $S' = \{x \in \mathcal{K}_1 : A'(x) - b' \in \mathcal{K}'_2\}$ we call $p'$ a $\mathcal{K}'_2$-relaxation. Moreover, for $\mathcal{K}'_2 = \mathbb{R}^m_{\geq 0}$ we call $p'$ a *linear relaxation* and for $\mathcal{K}_2 = S^m_{\geq 0}$, $p'$ forms a *semidefinite relaxation*.

Since the presentation of an optimization problem is not unique, it is important to formulate them as conveniently as possible. A brief description often enhances the performance of algorithms, on the theoretical side *Minkowski sums* or *Minkowski differences* often simplify the presentation of proofs and theorems such as the upcoming Theorem 2.1.5:

**Definition 2.1.4.** *Let $V$ be a vector space, then we define the Minkowski sum of two sets $A, B \subseteq V$ by*

$$A + B := \{a + b : a \in A, b \in B\}.$$

*Similarly, we define the Minkowski difference by*

$$A - B := \{c \in V : c + B \subseteq A\}.$$

We want to remark that balls $B(x, r) \subseteq \mathbb{R}^n$, often form Minkowski sums:

$$B(x, r) = B(x, r - \varepsilon) + B(x, \varepsilon), \qquad B(x, r - \varepsilon) = B(x, r) - B(x, \varepsilon).$$

They are subject of an application of Theorem 1.4.3 in Section 4.3.2.

## 2.1.1   Semidefinite programming and its complexity

The focus of this section is on the computational properties of semidefinite programs, where $\mathcal{K}_1 = S^n_{\geq 0}$ and $\mathcal{K}_2 = \{0\} \subseteq \mathbb{R}^m$ are the corresponding cones. The primal program then reads

$$SDP = \inf_{X \in S^n_{\geq 0}} \left\{ \langle C, X \rangle : \langle A_j, X \rangle = b_j \text{ for every } j \in [m] \right\}.$$

A common application of semidefinite programming, often abbreviated as SDP, is the approximation of 0/1-integer problems. Probably the most prominent such technique is the *Goemans-Williamson algorithm* [37] to tackle the MAX CUT problem on an undirected graph $G = (V, E)$ with edge weights $w_e = w_{uv} = w_{vu} \geq 0$:

$$\max \left\{ \frac{1}{4} \sum_{u,v \in V} w_{uv}(1 - x_u x_v) : x_v^2 = 1, \ v \in V \right\}.$$

The algorithm delivers an approximate solution of at least $0.87856\ldots$ times the optimal value.

Other applications include the *theta number* to approximate INDEPENDENT SET, i.e., the maximum number of vertices in a graph $G$, such that no two vertices are connected, and the GRAPH COLORING problem: Here we assign a color to every vertex in a graph such that no two connected vertices have the same color, these assignments are called *colorings*. The goal is to achieve a coloring that needs the fewest number of colors.

We dedicate Chapter 3 to semidefinite approximations of the SET COVER problem described in Chapter 1.

In general, semidefinite programs are not efficiently solvable in theory and practice as can be seen by, e.g., counterexamples of O'Donnell [62]. However, it is true, that under most circumstances semidefinite programs are in fact solvable in arbitrary precision within polynomial time – a result proven by Grötschel, Lovász and Schrijver [39] via the ellipsoid method and later also by de Klerk and Vallentin [20] via the interior point method. Both of them proved the following theorem.

**Theorem 2.1.5** (Grötschel, Lovász, Schrijver [39]). *Consider the semidefinite program*

$$SDP = \inf_{X \in S^n_{\geq 0}} \left\{ \langle C, X \rangle : \langle A_j, X \rangle = b_j \text{ for every } j \in [m] \right\},$$

*with rational input* $C, A_1, \ldots, A_m,$ *and* $b_1, \ldots, b_m.$ *Denote by*

$$\mathcal{F} = \left\{ X \in S^n_{\geq 0} : \langle A_j, X \rangle = b_j \text{ for every } j \in [m] \right\}$$

*the set of feasible solutions. Suppose we know a rational point* $X_0 \in \mathcal{F}$ *and positive rational numbers* $r, R$ *so that*

$$B(X_0, r) \subseteq \mathcal{F} \subseteq B(X_0, R),$$

*where $B(X_0, r)$ is the ball of radius r, centered at $X_0$, in the subspace*

$$L = \left\{ X \in \mathcal{S}^n \,:\, \langle A_j, X \rangle = 0 \text{ for every } j \in [m] \right\}.$$

*For every rational number $\varepsilon > 0$ one can find in polynomial time a rational matrix $X^* \in \mathcal{F}$ such that*

$$\langle C, X^* \rangle - \mathcal{SDP} \leq \varepsilon,$$

*where the polynomial is in n, m, $\log_2 \frac{R}{r}$, $\log_2 \frac{1}{\varepsilon}$ and the bit size of the data $X_0$, $A_1$, ..., $A_m$, and $b_1$, ..., $b_m$.*

For a fixed $\varepsilon > 0$ the problem of finding a rational matrix $X^*$ with the above properties or asserting that there is no such matrix with $B(X^*, \varepsilon) \subseteq \mathcal{F}$ is known as the *weak optimization problem*. Theorem 2.1.5 gives a sufficient criterion for polynomial computability of a wide class of SDPs. However, sometimes we need to deal with problems, where the interior is not full-dimensional or the set of feasible solutions $\mathcal{F}$ is not contained in a ball $B(X_0, R)$. In those cases the weak optimization problem can be proven to be solvable in polynomial time if the corresponding *weak separation problem* can be solved in polynomial time (see Corollary 4.2.7 in [41]).

**Definition 2.1.6** (Weak separation problem [41]). *Given a vector $y \in \mathbb{Q}^n$ and a rational number $\varepsilon > 0$, either*

- *assert that $y \in \mathcal{F} + B(0, \varepsilon)$, or*

- *find a vector $c \in \mathbb{Q}^n$ with $\|c\|_\infty = 1$ such that $c^T x \leq c^T y + \varepsilon$ for every x with $B(x, \varepsilon) \subseteq \mathcal{F}$.*

Here the sum $\mathcal{F} + B(0, \varepsilon)$ is again the Minkowski sum, which is for small values of $\varepsilon > 0$ a convenient tool to describe that $y$ is "almost in $\mathcal{F}$".
We demonstrate how to prove polynomial time solvability of a weak optimization problem via the weak separation problem in the following example. Our proof is inspired by a proof by Grötschel, Lovász and Schrijver [39] on the maximum weight of an independent set in a perfect graph.
One might wonder whether a simpler choice of indices for the following matrix $M$ would simplify the example below. In fact this is true, but the notation below facilitates the reasoning leading to the fact that the Lasserre hierarchy described in Chapter 3 is computable in polynomial time.

**Example 2.1.7.** *We first require a rational matrix $M \in \mathcal{S}^{(m+2) \cdot t \cdot n^t}$ to have the form*

$$M = \begin{pmatrix} M_t & 0 & \cdots & 0 \\ 0 & M_t^1 & \ddots & \vdots \\ \vdots & \ddots & \ddots & 0 \\ 0 & \cdots & 0 & M_t^{m+1} \end{pmatrix}$$

*and consider the corresponding set*

$$\mathcal{F} = \left\{ M \in \mathcal{S}_{\geq 0}^{(m+2) \cdot t \cdot n^t} : \langle A_j, M \rangle = b_j \text{ for every } j \in [T] \right\},$$

*where the constraints include setting the off-diagonal entries to 0. Suppose the columns with indices $b_1, \ldots, b_k$ of M form a basis of the subspace spanned by all columns. Then we consider the submatrix $M' \in \mathcal{S}^k$ of M that consists only of the entries $M_{ij}$ indexed by $i, j \in \{b_1, \ldots, b_k\}$. We observe that M is positive semidefinite if and only if $M'$ is positive definite. Thus we can check whether $M'$ is contained in $\mathcal{F}$ in polynomial time since it suffices to check whether the $(m + 2) \cdot t \cdot n^t$ determinants*

$$\det\left( M_l' \right) = \det\left( M_{ij}' \right)_{i, j \in [l]}$$

*for $l \in [k]$ are positive, this is known as Sylvester's criterion. Since the entries of $M'$ are rational we can compute the determinants in polynomial time (see [74]).*

*If M is not positive semidefinite then we have an index l being the smallest index such that $\det\left( M_l' \right) \leq 0$ holds. For $i \in [l]$ let $c_i = (-1)^i \cdot \tilde{M}_{i,l}'$, where $\tilde{M}_{i,l}'$ is the $(i, l)$-th minor of $M_l'$ and set $c_i = 0$ if $l < i < k$. After reindexing this yields $\langle cc^T, X \rangle \geq 0$ for every $X \in \mathcal{F}$ since $X \geq 0$ but*

$$\langle cc^T, M \rangle = \det M_l' \cdot \det M_{l-1}' \leq 0$$

*by definition. Thus the matrix $cc^T$ defines an almost separating hyperplane.*

## 2.1.2   The duality caused by the Riesz representation theorem

For the remainder of this section we will focus on a different cone, the cone of nonnegative Radon measures and define a convenient duality. Let $X$ be a compact topological Hausdorff space. Then we say that a measure $\mu$ on its $\sigma$-algebra of Borel sets $\mathcal{B}(X)$ is

- *inner regular* if, for every Borel set $B \in \mathcal{B}(X)$, $\mu(B) = \sup_{K \subseteq B, K \text{ compact}} \mu(K)$.

- *locally finite* if, for every compact Borel set $B' \in \mathcal{B}(X)$, $\mu(B') < \infty$.

**Definition 2.1.8** (Radon measure). *A Borel measure $\mu$ on a compact metric space $X$ is called a* Radon measure *if it is inner regular and locally finite. The set of Radon measures on $X$ is denoted by $\mathcal{M}(X)$.*

Sometimes in the literature these measures are referred to as *inner Radon measures* in contrast to *outer Radon measures* but in the case of a compact topological space $X$ these measures coincide. For more details on Radon measures we refer to Appendix B.2 in the book of Deitmar and Echterhoff [24] and to the book of Stein and Shakarchi [77]. In the spaces considered in this thesis we usually don't have to distinguish between Radon and Borel measures, since all of them are *Radon spaces*.

**Definition 2.1.9** (Radon space). *[2] A separable metric space X is a Radon space if every Borel probability measure is inner regular.*

Moreover, if $X$ is complete, it is a Radon space. To determine the duality, we again follow [24] to elaborate on the *Riesz representation theorem*.

**Theorem 2.1.10** (Riesz). *Let X be a compact metric space and denote the set of continuous real-valued functions on X by $C(X)$. Let $I : C(X) \to \mathbb{R}_{\geq 0}$ be a nonnegative linear functional. Then there exists a unique Radon measure $\mu : \mathcal{B}(X) \to [0, \infty]$ such that*

$$I(f) = \int_X f d\mu.$$

Thus we have linearity (continuity) of $I(f)$ for every function $f$ and Radon measure $\mu$ and further $I(f)$ is uniquely defined by $\mu$ (via Riesz representation theorem) and $f$ (since $I(f) = I(g)$ implies $f = g$ on the open sets in the *weak topology of the duality*, see [8]). This implies that the map

$$\langle \cdot, \cdot \rangle : \mathcal{M}(X) \times C(X) \to \mathbb{R}, \ \langle \mu, f \rangle := \int f d\mu$$

forms a duality. We observe that considering the cone $\mathcal{M}(X)_{\geq 0}$ and its dual $C(X)_{\geq 0}$ leads to conic programs, where we can apply the theory above.

In the next chapters of this thesis it is going to be necessary to show that if a (maybe not continuous) function $g$ satisfies a certain condition, then there also exists a continuous function $f$ satisfying this condition. To prove statements like this, the following lemma of Urysohn is a widely used tool.

**Lemma 2.1.11** (Urysohn's lemma). *Let X be a compact Hausdorff space and $A, B \subseteq X$ be closed and disjoint. Then there exists a continuous function $f : X \to [0, 1]$ such that $f(a) = 0$ for all $a \in A$ and $f(b) = 1$ for all $b \in B$.*

Thus if we have an indicator function $g = \mathbb{1}_B$ on $X$, we can approximate it with a continuous function $f \in C(X)$. For non-indicator functions, Tietze formulated an even stronger statement.

**Theorem 2.1.12** (Tietze's extension theorem). *Let X be a locally compact Hausdorff space and $g : A \to \mathbb{R}$ is a continuous map from a closed subset A of X. Then there exists a continuous function $f : X \to \mathbb{R}$ with the property $f(a) = g(a)$ for every $a \in A$. Moreover if $g$ is bounded, $f$ can be chosen to be bounded as well.*

These theorems will be very helpful, since if we have an optimization problem over the cone of nonnegative continuous functions $C(X)_{\geq 0}$, non-continuous functions satisfying the constraints turn out to have a continuous and thus feasible counterpart with the same objective value. In Chapter 5 we use this fact to show finite convergence of the hierarchy $\mathcal{N}^t(X, r)$, i.e., to prove the last equality in (1.4.4).

## 2.2  Groups and isometries

Until now, we have considered coverings of the central object of our thesis – a compact metric space $X$. A major drawback arises from this setup, since compactness rules out a couple of interesting setups such as coverings of the Euclidean space. In this section we address this problem by providing some background in group theory and we will further use this background in Section 2.3 to introduce techniques, that can be used to reduce the complexity of computing approximations of $N(X, r)$.

The principal example of this thesis is the spherical case $N(S^n, r)$. Every major result in Chapters 4 and 5 was carried out on the sphere first and later generalized to the broader setup of compact metric spaces. We can assign different groups to the sphere $S^n$, e.g., the *orthogonal group* $O(n + 1)$

$$O(n + 1) := \{A \in \mathbb{R}^{(n+1)\times(n+1)} : A^T A = I_{n+1}\}$$

or the corresponding *stabilizer subgroup* $\text{Stab}(O(n + 1), e)$ with respect to a point $e \in S^n$

$$\text{Stab}(O(n + 1), e) := \{\gamma \in O(n + 1) : \gamma e = e\}.$$

Note that $\text{Stab}(O(n + 1), e)$ and $O(n)$ are isomorphic. The two groups have the interesting property that they consist of *isometries* of the sphere with respect to the metric $d : S^n \times S^n \to \mathbb{R}$, $d(x, y) = \arccos(x \cdot y)$, that is for every $\tau \in O(n + 1)$ or $\tau \in \text{Stab}(O(n + 1), e)$ we have:

$$d(\tau x, \tau y) = d(x, y) \text{ for every } x, y \in S^n.$$

If we consider the set of all isometries $\tau : X \to X$, satisfying $d(\tau x, \tau y) = d(x, y)$ for every $x, y \in X$, acting on a compact metric space $X$, this also forms a group, referred to as the *isometry group* $\text{Iso}(X)$ of $X$. The isometry group will play a major role in Chapters 4 and 5.

An important concept to extend knowledge about coverings on compact metric spaces to non-compact metric spaces, such as the Euclidean space, is that we look at a compact subset of $\mathbb{R}^n$, cover this subset and then tesselate the whole Euclidean space. For this we need to find a suitable subset of $\mathbb{R}^n$, which can be done by considering the *quotient space* of $\mathbb{R}^n$ with respect to a specific group.

**Definition 2.2.1** (Quotient space). *Let $X$ be a topological space and $\sim$ be an equivalence relation. Then we define the quotient space $X/\sim$ as the set of equivalence classes with respect to $\sim$:*

$$X/\sim := \{[x] : x \in X\} := \{\{y \in X : y \sim x\} : x \in X\}.$$

We denote the map $\pi : X \to X/\sim$, $\pi(x) := [x]$ as the *canonical projection*. If the equivalence class is defined through a group action of a group $G$ acting on $X$, i.e., $x \sim y \Leftrightarrow \exists g \in G : x = gy$, we further denote the quotient space by $X/G$. The quotient space can be equipped with the following topology.

**Definition 2.2.2** (Quotient topology). *Let $(X, \mathcal{T})$ be a topological space with $\mathcal{T}$ the set of open sets, and $X/\!\!\sim$ its quotient space. Then we say that a set $U \subseteq X/\!\!\sim$ of equivalence classes is open in its* quotient topology, *if the union of its elements is open in $X$, i.e., $\bigcup_{[u] \in U} [u] \in \mathcal{T}$.*

This topology lets $(X, \mathcal{T})$ inherit a bunch of useful properties such as compactness and connectedness (see [46]). Observe that the integer numbers $\mathbb{Z}$ and also lattices introduced in Section 1.1 form a group. Thus, instead of covering the full Euclidean space we can instead cover the *flat torus*, a compact subset of $\mathbb{R}^n$, defined by $\mathbb{T}^n := \mathbb{R}^n/\mathbb{Z}^n$ or any automorphisms of it.

**Example 2.2.3** (Flat Torus). *A covering of the torus $\mathbb{T} := 2S^1 \times S^1 = \mathbb{R}^2/\left(\mathbb{Z}(2,0) + \mathbb{Z}(0,1)\right)$ can be illustrated by:*

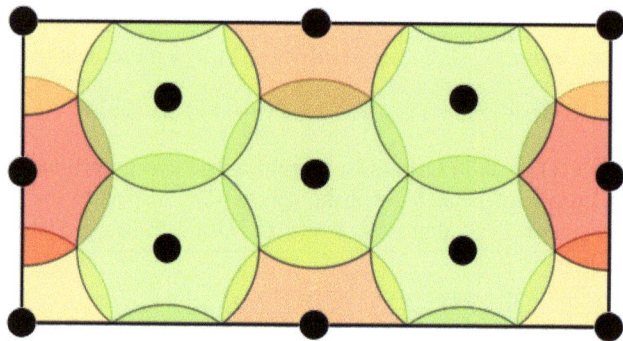

Figure 2.1: Flat torus covered by eight balls: 5 green balls and one red, orange and yellow ball respectively.

*We can tessellate the whole two dimensional Euclidean space by concatenating copies of such tori on the boundaries of each torus. The corresponding arrangement of balls provides a covering of Euclidean two-space with density*

$$8 \cdot \frac{\lambda(B(x, 5/16))}{\lambda(\mathbb{T})} = 8 \cdot \frac{\pi \cdot 5^2/16^2}{2} = \frac{25\pi}{64} \approx 1.227\ldots.$$

In the above example one point in our group corresponds to a whole torus. Thus the group acting on $\mathbb{R}^n$ is discrete and the benefit of considering the quotient space consists of the compactness of the corresponding quotient space.

The fact that a specific group action can sometimes extend covering results to non-compact spaces is important, but moreover, if the underlying metric space is compact, the following property is significant for the results in Chapter 5.

**Definition 2.2.4.** *We say that a group $\Gamma$ acts transitively on a space $X$ if and only if for every pair of elements $x, y \in X$ we have an element $\gamma \in \Gamma$ such that $\gamma x = y$ holds.*

A transitive group acting on a metric space $X$ has some remarkable properties. Among others, the quotient space consists of a single equivalence class and we can often translate between the group $\Gamma$ and $X$. This translation property will become important in Section 5.2.3.

If the isometry group $\mathrm{Iso}(X)$ is transitive we call the corresponding space $X$ *homogeneous*. In general, a compact metric space is not homogeneous. One class of counterexamples include the following situation: take a non-discrete metric space $X$ with diameter $< 1$ and add a point $x_0$ with distance 1 to every other point $x \in X \setminus \{x_0\}$. Then for $y \neq \tau^{-1} x_0$ we have $1 = d(x_0, y) = d(\tau x_0, \tau y) < 1$. Therefore, there is no transitive isometry in this case.

Nevertheless, a couple of important groups fall into this category such as the orthogonal group on the sphere, the projective orthogonal group on the projective space or the Euclidean group in $\mathbb{R}^n$. If we further consider compact metric spaces, it is remarkable that the corresponding isometry group can be equipped with a metric, and subsequently forms a compact metric space as well. This fact will be exploited in Section 4.3.3 and we prove it over the course of the following paragraphs as a consequence of the Arzelà-Ascoli Theorem 2.2.6. We recall the definition of equicontinuity first.

**Definition 2.2.5** (Equicontinuity). *Let $X$ be a compact metric space and $Y$ be a metric space, then a set $F \subseteq C(X, Y)$ is equicontinuous if for every $x \in X$ and $\varepsilon > 0$ $x$ has a neighbourhood $U_x$ such that*

$$d(f(y), f(x)) < \varepsilon \text{ for all } y \in U_x, f \in F.$$

The Arzelà-Ascoli Theorem is stated in different variations in the literature. We provide an adjusted version to the one of Kelley (see Theorems 7.21, 7.22 and 7.23 in [50]), but also the book of Folland [33] is a good source for further reading on this topic.

**Theorem 2.2.6** (Arzelà-Ascoli). *Let $X$ be a compact metric space and $Y$ be a metric space. Then $F \subseteq C(X, Y)$ is compact in the compact-open topology if and only if it is equicontinuous, pointwise relatively compact and closed.*

For compact metric spaces $X$ and $Y$ it is important to note that the compact open topology and the topology of pointwise convergence coincide (see Theorems 7.11 and 7.15 in Kelley [50]) and thus $\mathrm{Iso}(X)$ is metrizable with metric $e(\varphi, \tau) := \sup_{x \in X} \{d(\varphi(x), \tau(x))\}$ (see Proposition A.13 in [44]). Therefore, Theorem 2.2.6 provides a sufficient criterion to prove the following lemma that will be of use in Section 4.3.3. It is stated as Exercise 7.T.(a) in [50].

**Lemma 2.2.7.** *Let $\mathrm{Iso}(X)$ be the isometry group on a compact metric space $(X, d)$, then $(\mathrm{Iso}(X), e)$ is compact.*

*Proof.* First, we observe that the compact-open topology is metrizable with metric $e$. Let $\tau \in \mathrm{Iso}(X)$ be an arbitrary isometry. We show equicontinuity by the isometry property of $\tau$, that implies $d(\tau(x), \tau(y)) = d(x, y) < \delta := \varepsilon$ for any $\varepsilon > 0$.

For the pointwise relative compactness we consider the closure of $\mathrm{Iso}(X)_z := \{\tau(z) : \tau \in \mathrm{Iso}(X)\} \subseteq X$. As a subset of the compact metric space $X$ we know that it is totally bounded. Furthermore, as a closed subset of a complete (due to its compactness) metric space it is complete. Together we have that the closure of $\mathrm{Iso}(X)_z$ is compact and thus $\mathrm{Iso}(X)_z$ is pointwise relatively compact by definition.

Next we show that $(\mathrm{Iso}(X), e)$ is complete. Suppose we have a Cauchy sequence $\tau_k \in \mathrm{Iso}(X)$ then we know from Theorems 7.11, 7.12 and 7.13 in [50] that this Cauchy sequence has a limit $\tau$ in the set of continuous functions equipped with the compact open topology. Proposition A.13 in [44] then shows that the compact open topology is induced by the metric $e$ and thus the limit $\tau$ is continuous with respect to $e$.

We now show that the continuous function $\tau$ is an isometry. For sufficiently large $k \in \mathbb{N}$ we know that $e(\tau_k, \tau) < \varepsilon$ for any fixed $\varepsilon > 0$. This leads to

$$d(\tau(v), \tau(u)) \le d(\tau(v), \tau_k(v)) + d(\tau_k(v), \tau_k(u)) + d(\tau_k(u), \tau(u))$$
$$\le \sup_{x \in X} d(\tau(x), \tau_k(x)) + d(\tau_k(v), \tau_k(u)) + \sup_{x \in X} d(\tau_k(x), \tau(x))$$
$$< \varepsilon + d(v, u) + \varepsilon$$

on the one hand and to

$$d(v, u) = d(\tau_k(v), \tau_k(u))$$
$$\le d(\tau_k(v), \tau(v)) + d(\tau(v), \tau(u)) + d(\tau(u), \tau_k(u))$$
$$\le \sup_{x \in X} d(\tau_k(x), \tau(x)) + d(\tau(v), \tau(u)) + \sup_{x \in X} d(\tau(x), \tau_k(x))$$
$$< \varepsilon + d(\tau(v), \tau(u)) + \varepsilon$$

on the other hand. Since those inequalities hold for any fixed $\varepsilon > 0$, we have that $d(v, u) = d(\tau(v), \tau(u))$ for any $u, v \in X$ making $\tau$ an isometry.

This isometry is also invertible:

$$\sup_{x \in X} d(\tau_k^{-1} \tau(x), x) = \sup_{x \in X} d(\tau(x), \tau_k(x)) \to 0.$$

Thus $\tau_k^{-1} \to \tau^{-1}$ implying that $\tau$ is bijective. Therefore $(\mathrm{Iso}(X), e)$ is a complete metric space and thus closed. Finally we fulfill each condition to apply the Theorem of Arzelà-Ascoli and finish the proof.                                                                                  □

The shown compactness plays a major role in extending the main result in Chapter 4 on coverings of $\mathbb{R}^n$ by Euclidean balls, to coverings of $\mathbb{R}^n$ by translates of finite unions of Euclidean balls. Since all measurable bodies with non-empty interior can be approximated quite well with unions of balls this allows us to generalize Theorem 4.1.1 in Euclidean space.

## 2.3   Symmetry reduction and harmonic analysis

The final section of this chapter is dedicated to the concept of symmetry reduction for semidefinite programs. These techniques are well-established and have been successfully applied to a variety of problems, see for example [6], [35] or [65]. Additionally, we will apply them to approximate the covering number $N(S^2, r)$ in Section 5.7. In the present section we mainly follow Vallentin [80] to explain the necessary methods and techniques for Section 5.7.

We introduce the technique of symmetry reduction on metric spaces $(X, d, \omega)$ equipped with metric $d$ and finite, regular Borel measure $\omega$. In particular, we know from Section 2.1.2 that $\omega$ is consequently a Radon measure. The group $\Gamma$ over which we symmetrize is supposed to be a compact, transitive subgroup of the isometry group $\mathrm{Iso}(X)$. Here for symmetric kernels $K \in C(X \times X)$ the *shifted kernels* $K^\gamma$ for $\gamma \in \Gamma$ are defined by $K^\gamma(x, y) = K(\gamma x, \gamma y)$. For defining symmetrized kernels one needs the Haar measure of a group $\Gamma$:

**Definition 2.3.1** (Haar measure). *Let $\Gamma$ be a locally compact group. Then a (left) Haar measure $\lambda$ is a left invariant regular Borel measure, that is positive on any non-empty, open Borel set $B \subseteq \Gamma$. We call $\lambda$ left invariant if and only if*

$$\lambda(\gamma B) = \lambda(B) \text{ for every Borel set } B \text{ and every } \gamma \in \Gamma.$$

The proof of the existence of such a Haar measure can be found for example in [24]. Let us denote by $C(X \times X)^\Gamma$ the set of $\Gamma$-*invariant kernels* consisting of kernels $K$ satisfying $K = K^\gamma$ for every $\gamma \in \Gamma$. We further call such kernels positive semidefinite if for every finite subset $J \subseteq X$, the matrix $(K(i, j))_{i,j \in J}$ is positive semidefinite. We denote the corresponding cone by $C(X \times X)^\Gamma_{\geq 0}$. A simple example of such an invariant kernel is the integral over the shifted kernels with respect to a normalized Haar measure on $\Gamma$:

$$\int_\Gamma K^\gamma d\lambda(\gamma) \in C(X \times X)^\Gamma_{\geq 0}.$$

A benefit of this $\Gamma$-invariant kernel stems from $\Gamma$-*invariant conic programs* over the cone $C(X \times X)_{\geq 0}$, i.e., conic programs for which a feasible solution $K \in C(X \times X)_{\geq 0}$ yields feasible solutions $K^\gamma \in C(X \times X)_{\geq 0}$ for every $\gamma \in \Gamma$. The above integral is then also a feasible solution of the program and thus, if the objective values of $K$ and $K^\gamma$ coincide, we can restrict the program to the cone $C(X \times X)^\Gamma_{\geq 0}$.

In the next paragraphs we want to introduce some techniques to decompose cones of invariant kernels into smaller cones. For this, we first define an *orthonormal system* $e_1, e_2, \ldots \in C(X)$ by

$$\int_X e_k(x)e_k(x)d\omega(x) = 1, \text{ and } \int_X e_k(x)e_l(x)d\omega(x) = 0, \text{ whenever } k \neq l.$$

Such a system is *complete* if every continuous function $f \in C(X)$ can be approximated arbitrarily well by finite linear combinations of $e_k$ in terms of convergence with respect to the $L^2$-norm :

$$\|f\| := \sqrt{\int_X f^2(x)d\mu(x)}.$$

An example of such a system is a basis in the subalgebra of polynomials on the sphere $S^n$. This algebra can be used to approximate the algebra of continuous, real-valued functions on $S^n$, which is due to the *Stone-Weierstrass theorem*.

**Theorem 2.3.2** (Stone-Weierstrass [75]). *Let $X$ be a compact Hausdorff space. Let $\mathcal{A}$ be a subalgebra of $C(X)$, so that for any $x, y \in X$ and $\alpha, \beta \in \mathbb{R}$, there exists $f \in \mathcal{A}$ with $f(x) = \alpha$ and $f(y) = \beta$. Then $\mathcal{A}$ is dense in $C(X)$ with respect to $\| \cdot \|_{\infty}$.*

The approximation of continuous functions on the sphere by polynomials will be addressed again in Section 5.7.2. Another, more general, way of constructing an orthonormal system is given by the proof in [80] of the following Theorem of Peter and Weyl (see [63] and [83]). Here we first need to define a couple of objects for a transitive group $\Gamma \subseteq \text{Iso}(X)$:

- A subspace $S \subseteq C(X)$ is called $\Gamma$-*invariant* if $\gamma S = S$ for every $\gamma \in \Gamma$.

- A nonzero subspace $S$ is called $\Gamma$-*irreducible* if $\{0\}$ and $S$ are the only $\Gamma$-invariant subspaces of $S$.

- For two $\Gamma$-invariant subspaces $F, F' \subseteq C(X)$ a linear map $T : F \to F'$ is called a $\Gamma$-*map* if $T(\gamma f) = \gamma T(f)$.

- We say that two spaces $S, S' \subseteq C(X)$ are $\Gamma$-*equivalent* if there is a bijective $\Gamma$-map between them.

We can now state the theorem of Peter and Weyl.

**Theorem 2.3.3** (Peter-Weyl). *All $\Gamma$-irreducible subspaces of $C(X)$ are of finite dimension. The space $C(X)$ decomposes orthogonally as*

$$C(X) = \bigoplus_{k=0}^{\infty} H_k,$$

*and the space $H_k$ decomposes orthogonally as*

$$H_k = \bigoplus_{i=0}^{m_k} H_{k,i},$$

*where $H_{k,i}$ is irreducible, and $H_{k,i}$ is equivalent to $H_{k',i}$ if and only if $k = k'$. The dimension $h_k$ of $H_{k,i}$ is finite, but the multiplicity $m_k$ can potentially be infinite.*
*In other words, $C(X)$ has a complete orthonormal system $e_{k,i,l}$, where $k = 0, 1, \ldots$, the index $i$ ranges from 1 to $m_k$, where $m_k$ is potentially infinite, $l = 1, 2, \ldots, h_k$, $h_k$ finite, so that*

(1) the space $H_{k,i}$ spanned by $e_{k,i,1}, \ldots, e_{k,i,h_k}$ is irreducible,

(2) the spaces $H_{k,i}$ and $H_{k',i'}$ are equivalent if and only if $k = k'$,

(3) there are $\Gamma$-maps $\phi_{k,i} : H_{k,1} \rightarrow H_{k,i}$ mapping $e_{k,1,l}$ to $e_{k,i,l}$.

Orthonormal systems satisfying properties (1) to (3) are perfectly suited as an input to the following theorem of Bochner [10], a classical result in harmonic analysis.

**Theorem 2.3.4** (Bochner). *Let $e_{k,i,l}$ be a complete orthonormal system for $C(X)$ as in Theorem 2.3.3. Every $\Gamma$-invariant, positive semidefinite kernel $K \in C(X \times X)_{\geq 0}^{\Gamma}$ can be written as*

$$K(x,y) = \sum_{k=0}^{\infty} \sum_{i,j=1}^{m_k} f_{k,ij} \sum_{l=1}^{h_k} e_{k,i,l}(x) e_{k,j,l}(y), \tag{2.3}$$

*or more economically as*

$$K(x,y) = \sum_{k=0}^{\infty} \langle F_k, Z_k^{(x,y)} \rangle, \tag{2.4}$$

*with $(F_k)_{ij} = f_{k,ij}$ and $\left(Z_k^{(x,y)}\right)_{ij} = \sum_{l=1}^{h_k} e_{k,i,l}(x) e_{k,j,l}(y)$. Here $F_k$ is symmetric and positive semidefinite. If $\Gamma$ operates transitively on $X$, it has one orbit, the multiplicities $m_k < \infty$ are finite, and the series $\sum_{k=0}^{\infty} \langle F_k, Z_k^{(x,y)} \rangle$ converges absolutely-uniformly.*

If the group $\Gamma$ is not transitive, we turn to de Laat [22]. If $\Gamma$ has finitely many orbits, he proved the absolutely-uniform convergence of $\sum_{k=0}^{\infty} \langle F_k, Z_k^{(x,y)} \rangle$ in Theorem 3.4.4. For $\Gamma$ having infinitely many orbits, he constructed a sequence

$$C_1 \subseteq C_2 \subseteq \ldots \subseteq C(X \times X)_{\geq 0}^{\Gamma},$$

whose union $\bigcup_{d=1}^{\infty} C_d$ is dense in $C(X \times X)_{\geq 0}^{\Gamma}$. We will elaborate on this sequence in Section 5.7.1.

We demonstrate the application of Bochner's theorem in an exemplary manner by considering two different optimization problems. First, we consider a finite semidefinite program:

$$\inf_{K \in S_{\geq 0}^n} \left\{ \langle C, K \rangle : \langle A_j, K \rangle - b_j = 0 \text{ for every } j \in [m] \right\} \tag{2.5}$$

and later extend this technique to a conic program over the cone $C(S^n \times S^n)_{\geq 0}$. For (2.5), we consider the metric space $([n], d, \omega)$, where the metric

$$d(j,k) := \arccos\left( \begin{pmatrix} \sin \frac{2\pi j}{n} \\ \cos \frac{2\pi j}{n} \end{pmatrix} \cdot \begin{pmatrix} \sin \frac{2\pi k}{n} \\ \cos \frac{2\pi k}{n} \end{pmatrix} \right),$$

measures the standard distance of the numbers $1, \ldots, n$ projected on a circle. For $n = 11$ we can essentially picture a clock and $d$ measures the shortest path on the boundary between two times. In this example we let the group $\Gamma \subset \text{Iso}([n])$ be the *cyclic group*, consisting of rotations of multiples of $\frac{2\pi}{n}$ on this circle. This group is transitive. We define the measure $\omega$ by $\omega(A) = |A|$.

Suppose that (2.5) is invariant under $\Gamma$, i.e., each feasible solution $K$ yields feasible solutions $K^\gamma$ defined by $K^\gamma(x, y) = K(\gamma x, \gamma y)$ for every $\gamma \in \Gamma$ and the objective values of $K$ and $K^\gamma$ coincide. This leads to the crucial observation that also the *symmetrized* solution corresponding to $K$

$$(x, y) \mapsto \frac{1}{|\Gamma|} \sum_{\gamma \in \Gamma} K(\gamma x, \gamma y)$$

is a feasible solution to (2.5). Additionally, it has the same objective value and thus we can restrict the program (2.5) to $\Gamma$-invariant matrices $K$. We recall the notation of the cone of such matrices as $C([n] \times [n])_{\geq 0}^\Gamma$.

The Peter-Weyl theorem now tells us that there exists an orthonormal system $e_{k,i,l} \in C([n]) = \mathbb{R}^n$ such that the space $C([n])$ decomposes orthogonally as

$$C([n]) = \bigoplus_{k=0}^{N} \bigoplus_{i=0}^{m_k} H_{k,i},$$

to its irreducible subspaces $H_{k,i}$ spanned by $e_{k,i,1}, \ldots, e_{k,i,h_k}$. We can use this system as an input for Bochner's theorem and rewrite every $\Gamma$-invariant, positive semidefinite matrix $K \in C([n] \times [n])_{\geq 0}^\Gamma$ as

$$K(x, y) = \sum_{k=0}^{N} \sum_{i,j=1}^{m_k} f_{k,ij} \sum_{l=1}^{h_k} e_{k,i,l}(x) e_{k,j,l}(y), \tag{2.6}$$

or more economically as

$$K(x, y) = \sum_{k=0}^{N} \langle F_k, Z_k^{(x,y)} \rangle, \tag{2.7}$$

with $(F_k)_{ij} = f_{k,ij}$ and $\left( Z_k^{(x,y)} \right)_{ij} = \sum_{l=1}^{h_k} e_{k,i,l}(x) e_{k,j,l}(y)$. Here $F_k$ is symmetric and positive semidefinite.

One observes that the parameters $N \leq n, m_1, \ldots, m_N$ and every parameter to define the matrices $Z_k$ depend on the group $\Gamma$. The benefit of this decomposition is the following reformulation of the semidefinite program (2.5):

$$\inf \langle C, K \rangle$$

$$\left\langle A_j, \sum_{k=0}^{N} \langle F_k, Z_k \rangle \right\rangle - b_j = 0 \text{ for every } j \in [m]$$

$$F_1, \ldots, F_N \in S_{\geq 0}^{m_1}, \ldots, S_{\geq 0}^{m_N}.$$

Computationally this often turns out to have two major advantages. First, for many practical solvers the block structure speeds up the numerical calculations and second, the sum $m_1 + \ldots + m_N$ is in many applications much smaller than the dimension $n$ of our original problem reducing the size of the given instance significantly.

A second example of the power of block diagonalization introduces some necessary objects to illustrate the techniques used in Section 5.7. The conic program we consider is the following

$$\sup\left\{\int_{S^n \times S^n} K d\mu \; : \; C - B(K) = 0, \; K \in C(S^n \times S^n)^{O(n+1)}_{\geq 0}\right\},$$

where $\mu \in \mathcal{M}(S^n \times S^n)_{sym.}$ is a symmetric Radon measure, $C \in \mathbb{R}$ and $B : C(S^n \times S^n) \to \mathbb{R}$ is a linear operator. Before we apply Theorem 2.3.4 we consider some results of Schoenberg [73], stating that the multiplicities satisfy $m_k = 1$ and the one-dimensional kernels $Z_k$ are determined by normalized *Jacobi polynomials* $P_k^{(\alpha,\alpha)}$ of degree $k$ with parameters $(\alpha, \alpha)$ and $\alpha = \frac{n-2}{2}$, i.e., $Z_k^{(x,y)} = P_k^{(\alpha,\alpha)}(x \cdot y)$. Consequently, Theorem 2.3.4 leads to a linear program

$$\sup\left\{\sum_{k=0}^{\infty} f_k \int_{S^n \times S^n} P_k^{(\alpha,\alpha)}(x \cdot y) d\mu(x,y) \; : \; C - \sum_{k=0}^{\infty} f_k B\left(P_k^{(\alpha,\alpha)}\right) = 0, \; f_k \in \mathbb{R}_{\geq 0}\right\}. \tag{2.8}$$

The convergence of the sum $\sum_{k=0}^{\infty} f_k P_k^{(\alpha,\alpha)}(x \cdot y)$ is a consequence of the fact that the Jacobi polynomials $P_1^{(\alpha,\alpha)}, P_2^{(\alpha,\alpha)}, \ldots$ form a complete orthogonal system of $L^2([-1,1],(1-t^2)^{\alpha})$. The approach to approximate the covering number $N(S^2, r)$ in Section 5.7.2 falls into a similar setup. Here we consider in addition the stabilizer subgroup $\mathrm{Stab}(O(n+1), e)$ with respect to a fixed point $e \in S^n$. Although this group is not transitive, Bachoc and Vallentin [6] provided a similar characterization of $C(S^n \times S^n)_{\geq 0}^{\mathrm{Stab}(O(n+1),e)}$ using *Gegenbauer polynomials* $C_k^{\lambda}$ with parameter $\lambda = \frac{n}{2} - 1$. Gegenbauer polynomials can be defined inductively by $C_0^{\lambda}(t) := 1$, $C_1^{\lambda}(t) := 2\lambda t$ and

$$C_k^{\lambda}(t) := 2(k + \lambda - 1)t C_{k-1}^{\lambda}(t) - (k + 2\lambda - 2)C_{k-1}^{\lambda} \text{ for } k \geq 2.$$

For more details on Jacobi and Gegenbauer polynomials we refer to the book of Andrews, Askey and Roy [3].

# CHAPTER THREE

# The Lasserre hierarchy on SET COVER

This chapter is based on the Bachelor's thesis *Approximating Set Cover using the Lasserre hierarchy* written by T. Acisu [1].

We recall the definition of SET COVER, denoted by *SC*, as seen in Section 1.4 and give a few basic examples. Additionally, we consider the lower bound given by the linear relaxation of *SC* and use it to give a variant of the proof of Chvátal's [13] Theorem 3.1.6. In Chapter 4 we will extend this approach to compact metric spaces.

In Section 3.2 we first give some general properties of Lasserre's [54] famous hierarchy to approximate 0/1-integer programs and prove that it converges in finitely many steps, in particular that if we have covering sets $S_1, \ldots, S_n$, the $n$-th step in Lasserre's hierarchy and *SC* coincide. This should give a good intuition for the content upcoming in Chapter 5. Finally, we present a subexponential algorithm of Chlamtac, Friggstad and Georgiou [12] in a version of Rothvoß [71] that improves Chvátal's upper bound by using the Lasserre relaxations.

## 3.1 Bounds based on linear relaxations

### 3.1.1 Preliminaries and basic examples

We recall the definition of the problem SET COVER, where we denote $\{1, \ldots, m\}$ by $[m]$.

**Definition 3.1.1.** *Given a set $[m]$ as well as a collection of subsets $S_1, \ldots, S_n \subseteq [m]$ with $\bigcup_{i=1}^{n} S_i = [m]$ and assigned costs $c_i \in \mathbb{Q}_{>0}$ for each subset $S_i$. The goal is to find a set of indices $\mathcal{J}^* \subseteq [n]$, such that:*

$$\bigcup_{i \in \mathcal{J}^*} S_i = [m] \text{ and } \sum_{i \in \mathcal{J}^*} c_i \text{ is minimal.} \tag{3.1}$$

*The subcollection $\{S_i : i \in \mathcal{J}^*\}$ is defined as a minimum set cover.*

We simply call $\mathcal{J}$ a *cover* of $[m]$ if $\bigcup_{i \in \mathcal{J}} S_i = [m]$. Next, we give two basic examples.

**Example 3.1.2.** *Suppose we have the ground set* $\{1, \ldots, 14\}$ *and 5 subsets* $S_1, \ldots, S_5$ *defined by Figure 3.1 with assigned costs* $c_1 = \ldots = c_5 = 1$.

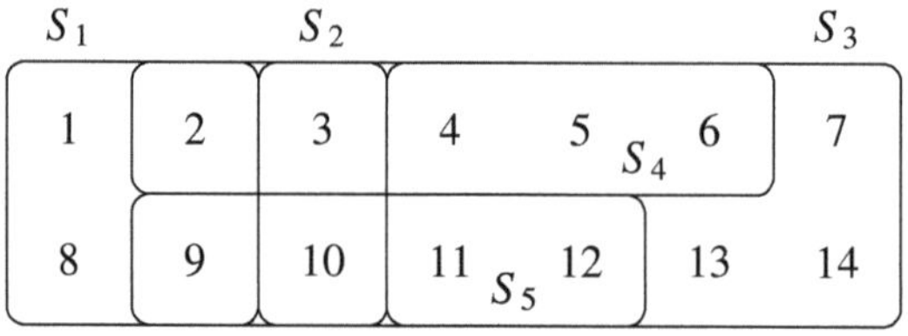

Figure 3.1: One instance of set cover with sets $S_1, \ldots, S_5$ and elements $1, \ldots, 14$

We observe that for each of the sets $S_1, S_3$ there is a point that is just covered by this set. Thus we need to include the indices 1 and 3 in our cover $\mathcal{J}^*$. The elements 3 and 10 can either be covered by adding $S_2$ and cost 1 or by adding any two sets out of $\{S_2, S_4, S_5\}$ with cost 2. Consequently we add $S_2$ and achieve an optimal cover $\mathcal{J}^* = \{1, 2, 3\}$ of cost 3.

**Example 3.1.3.** *We further present an example inspired by Haynes, Hedetniemi and Slater [45] and presented in a related matter in [69] by Rolfes. Suppose, in a remote part of the world like the Outback in Australia or Siberia, we want to locate radio stations in some of the very rare villages in these regions and ensure that each village receives the radio program despite the limited broadcasting range.*
*Since in this example radio stations are equally expensive, minimizing the total cost corresponds to minimizing the number of stations. Let the broadcasting range be 50 kilometres, then for every villlage* $v_i$, *we consider the sets* $S_i$ *of villages within this broadcasting range. The following graph connects two villages whenever they can broadcast to each other, but omits loops:*
    *Here the two sets* $S_5 = \{v_1, v_4, v_5, v_7, v_9\}$ *and* $S_6 = \{v_2, v_3, v_6, v_7, v_8\}$ *cover the whole ground set* $\{v_1, \ldots, v_9\}$.

In 1972, Karp [49] showed, that SET COVER is NP-complete. This means, if P $\neq$ NP, although the feasibility of a proposed solution can be verified in polynomial time, it is widely believed that the problem can not be solved in polynomial time. In fact, it was shown by Dinur and Steurer [27], that there is no efficient way of solving any given instance of SET COVER to within a factor of $(1 - \varepsilon) \ln m$. On the contrary, we do have methods to approximate a solution to a certain degree in a reasonable amount of time. To point out the relation of SET COVER to the field of combinatorial optimization and to be able to apply techniques from this field, one needs to reformulate SET COVER as an integer linear program:

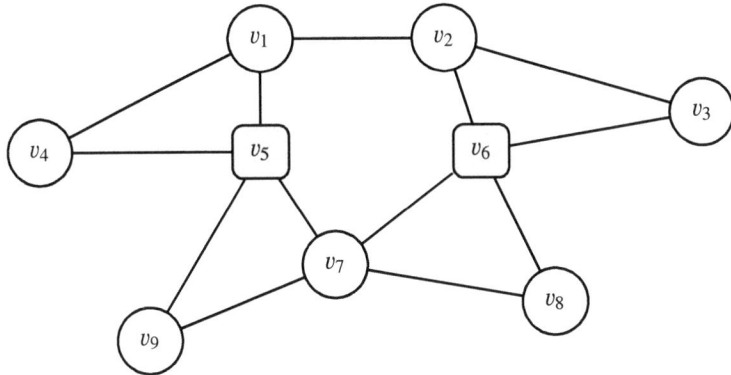

Figure 3.2: Two radio stations at villages $v_5, v_6$ broadcasting to 9 villages $v_1, ..., v_9$.

**Definition 3.1.4** (Integer linear program for SET COVER).

$$SC := \min\left\{\sum_{i=1}^{n} c_i x_i : x \in \{0, 1\}^n, \sum_{i:\, k\in S_i} x_i \geq 1 \text{ for every } k \in [m]\right\}. \tag{3.2}$$

We denote by $\mathcal{J}^* = \{i \in [n] : x_i^* = 1\}$ the support of an optimal solution $x^*$. Since integer linear programs are in general NP-hard, a first natural approximating step is weakening the integrality constraints on $x$ to achieve a linear program. These programs can be solved efficiently by using well-known methods such as interior point methods or the ellipsoid method.

**Definition 3.1.5** (Linear relaxation of $SC$).

$$SC \geq \min\left\{\sum_{i=1}^{n} c_i x_i : x \in [0, 1]^n, \sum_{i:\, k\in S_i} x_i \geq 1 \text{ for every } k \in [m]\right\}. \tag{3.3}$$

## 3.1.2 Greedy Algorithm

Even though finding an exact solution for SET COVER is difficult, as already mentioned in Section 1.4, there are logarithmic upper bounds known. A straightforward way to find such a solution is a greedy algorithm: We consider an empty index set $\mathcal{J}$ and look for the subset $S_i$ with the least cost per covered element $\frac{c_i}{|S_i|}$ and include it in the set $\mathcal{J}$. In the next step we repeat these actions but this time we only consider the elements, which are not yet included in the set $\bigcup_{i\in\mathcal{J}} S_i$. Finally, this process results in a cover $\mathcal{J}$ satisfying $\bigcup_{i\in\mathcal{J}} S_i = [m]$. Observe that instead of asking for $\frac{c_i}{|S_i|}$ being minimal, one could ask for $\frac{|S_i|}{c_i}$ being maximal, then formally the algorithm states

---

**Algorithm 1** Greedy Algorithm for SET COVER

---

1: $\mathcal{J} \leftarrow \emptyset$
2: $r \leftarrow 0$
3: $S_i^r \leftarrow S_i$ for all $i \in [n]$
4: **while** $\bigcup_{i \in \mathcal{J}} S_i \neq [m]$ **do**
5:    $r \leftarrow r + 1$
6:    Choose $j \in [n] \setminus \mathcal{J}$ with $\frac{|S_j^{r-1}|}{c_j} \geq \frac{|S_i^{r-1}|}{c_i}$ for all $i \in [n]$
7:    $\mathcal{J} \leftarrow \mathcal{J} \cup \{j\}$
8:    $S_i^r \leftarrow S_i^{r-1} \setminus S_j^{r-1}$ for all $i \in [n]$
9: **end while**

---

Since we chose the cheapest sets in terms of costs per element in every step, intuitively the returned cover should be rather cheap. In fact in 1979 Chvátal [13] analyzed this algorithm and reached the following result.

**Theorem 3.1.6.** *Let $M = \max_{i \in [n]} |S_i|$ being the size of the largest subset $S_i$ and $c$ be the cost of a cover $\mathcal{J}$ returned by the greedy algorithm. Then*

$$c \leq H(M) \cdot SC$$

*with $H(M)$ being the harmonic series $H(M) = \sum_{k=1}^{M} \frac{1}{k} \leq \ln M + 1$.*

The proof we provide here is slightly modified to point out the relation to the linear relaxation (3.3) but the proof still keeps the structure of Chvátal's [13] original proof.

*Proof.* First we observe that the algorithm returns a cover: In every iteration it covers at least one element that is not already covered and thus after at most $m$ iterations the algorithm terminates with a cover $\mathcal{J}$ satisfying $\bigcup_{i \in \mathcal{J}} S_i = [m]$. Thus it remains to show that the cover $\mathcal{J}$ has, in the worst case, a cost of $H(M)$ times the optimal cost. For this we examine the linear program (3.3), omitting the neglectable $x_i \leq 1$ constraints

$$\min \left\{ c^T x : \ x \geq 0, \ Ax \geq \mathbb{1} \right\}$$

and its corresponding dual

$$\max \left\{ \mathbb{1}^T y : \ y \geq 0, \ A^T y \leq c \right\},$$

where $A_{ki} = 1$ if $k \in S_i$ and 0 otherwise. The greedy algorithm will now provide a feasible solution $\frac{z}{H(M)} \in \mathbb{R}^m$ for the dual program by setting $z_k$ as the average cost that is paid for covering element $k$ by the greedy heuristic: Suppose the element $k$ is covered in the $r$-th step of the greedy algorithm by picking the index $j$ with $\frac{|S_j^{r-1}|}{c_j}$ maximal. Observe that $r$ and $k$ have a one-to-one relation since exactly after iteration $r$, the element $k$ is removed from the remaining sets, i.e., $k \notin S_i^r$. Let $w_j^r = |S_{,j}^r|$, then we define the average costs for $k \in S_{,j}^{r-1}$ being added to the cover in iteration $r$ by picking the set $S_{,j}^{r-1}$ as

$$z_k = \frac{c_j}{w_j^{r-1}} = \frac{\text{cost of the covering set } S_j}{\text{number of elements in } S_j \text{ after } r - 1 \text{ iterations}}.$$

We check the feasibility of the dual program: The constraint $z \geq 0$ is immediate since both numerator and denominator are positive. We further show that $z$ satisfies

$$(A^T z)_i \leq H(M) \cdot c_i$$

implying feasibility of $\frac{z}{H(M)}$. The key identity here is $S_i \cap S_j^{r-1} = S_i^{r-1} \setminus S_i^r$, since both sets contain exactly the elements of the original set $S_i$ that get covered in iteration $r$ by picking $S_j^{r-1}$. Suppose the algorithm has iterations $r$ ranging from 1 to $b$, i.e., $b$ is the largest iteration such that there is an index $k$ with $w_k^{b-1} > 0$. Then,

$$(A^T z)_i = \sum_{k \in S_i} z_k = \sum_{r=1}^{b} \sum_{k \in S_i \cap S_j^{r-1}} z_k = \sum_{r=1}^{b} \sum_{k \in S_i^{r-1} \setminus S_i^r} z_k = \sum_{r=1}^{b} (w_i^{r-1} - w_i^r) \frac{c_j}{w_j^{r-1}}.$$

Since in iteration $r$ we have $\frac{c_j}{w_j^{r-1}}$ chosen minimal, we obtain

$$\sum_{r=1}^{b} (w_i^{r-1} - w_i^r) \frac{c_j}{w_j^{r-1}} \leq c_i \sum_{r=1}^{b} \frac{w_i^{r-1} - w_i^r}{w_i^{r-1}}.$$

Finally we bound with respect to the harmonic series as follows:

$$c_i \sum_{r=1}^{b} (w_i^{r-1} - w_i^r) \frac{1}{w_i^{r-1}} = c_i \sum_{r=1}^{b} \left( \frac{1}{w_i^{r-1}} + \ldots + \frac{1}{w_i^{r-1}} \right)$$

$$\leq c_i \sum_{r=1}^{b} \left( \frac{1}{w_i^{r-1}} + \frac{1}{w_i^{r-1} - 1} + \ldots + \frac{1}{w_i^r + 1} \right)$$

$$= c_i \sum_{r=1}^{b} \left( \frac{1}{w_i^{r-1}} + \frac{1}{w_i^{r-1} - 1} + \ldots + \frac{1}{w_i^r + 1} + H(w_i^r) - H(w_i^r) \right)$$

$$= c_i \sum_{r=1}^{b} \left( H(w_i^{r-1}) - H(w_i^r) \right) = c_i \cdot H(w_i^0)$$

$$= c_i \cdot H(|S_i|) \leq c_i \cdot \max_{i \in [n]} H(|S_i|) = c_i \cdot H(M).$$

Thus we have $\left( A^T z \right)_i \leq c_i \cdot H(M)$ for every index $i \in [n]$ implying the dual feasibility of $\frac{z}{H(M)}$. If we compute the objective values of the vector $\frac{z}{H(M)}$ and of an optimal solution $x^*$

of the primal program (3.3) we obtain by weak duality:

$$\frac{1^T z}{H(M)} \le c^T x^* \le SC.$$

On the other hand, since the greedy algorithm provides us with a cover $\mathcal{J}$, the corresponding index vector $x' \in \{0, 1\}^n$ supported on $\mathcal{J}$ is feasible for the primal program and we obtain for the objective value

$$c^T x' = \sum_{j \in \mathcal{J}} c_j = \sum_{j \in \mathcal{J}} w_j^{r-1} \frac{c_j}{w_j^{r-1}} = \sum_{j \in \mathcal{J}} \sum_{k \in S_j^{r-1}} z_k = 1^T z.$$

Combining these two (in-)equalities finally leads to

$$c^T x' = 1^T z \le H(M) \cdot SC$$

and closes the proof.    □

Hence we have shown that the greedy algorithm yields an approximation ratio of

$$\frac{c^T x'}{SC} \le H(M) = \sum_{j=1}^{M} \frac{1}{j} = \ln(M) + \gamma + o\left(\frac{1}{2M}\right) \le \ln(M) + 1,$$

where $\gamma$ is known as the *Euler-Mascheroni constant*. This implies that a logarithmic upper bound for SET COVER can be computed in polynomial time. Furthermore, due to the result of Dinur and Steurer [27], unless P = NP, it is very unlikely that one can efficiently achieve a better approximation ratio. However, in subexponential time, it is possible to present a better approximating algorithm that is based on the famous hierarchy of Lasserre [54].

## 3.2    The Lasserre hierarchy

The Lasserre hierarchy was designed as a method to approximate general 0/1-integer problems, such as

$$\mathcal{IP} = \max\left\{c^T x : x \in \{0, 1\}^n, Ax \ge b\right\}, \tag{3.4}$$

where $c \in \mathbb{R}^n$, $A \in \mathbb{R}^{m \times n}$ and $b \in \mathbb{R}^m$. We relax the integrality constraints and consider the polyhedron

$$K = \left\{x \in \mathbb{R}^n : \begin{pmatrix} A \\ I \\ -I \end{pmatrix} x \ge \begin{pmatrix} b \\ 0 \\ -1 \end{pmatrix}\right\}.$$

To simplify the notation we will denote $A = \begin{pmatrix} A \\ I \\ -I \end{pmatrix}$ and $b = \begin{pmatrix} b \\ 0 \\ -1 \end{pmatrix}$ for the rest of this chapter.

The optimal value $\mathcal{IP}$ is attained at a vertex of the polytope $P = \text{conv}\{K \cap \{0, 1\}^n\}$, the convex hull of $K$ intersected with the vertices of the $n$-dimensional unit cube. What the Lasserre hierarchy achieves, is to construct a hierarchy of convex bodies $L_t(K)$ with respect to $K$, that converges to $P$ after at most $n$ steps:

$$K = L_0(K) \supseteq L_1(K) \supseteq \cdots \supseteq L_n(K) = P. \tag{3.5}$$

This leads to the definition of the central concept of this chapter, the Lasserre hierarchy of level $t$ (as defined in [71]), where the involved vectors are indexed by $\mathcal{I}_t := \{I \in \mathcal{P}([n]) : |I| \le t\}$.

**Definition 3.2.1.** *Let* $K = \{x \in \mathbb{R}^n : Ax \ge b\}$ *defined by* $A \in \mathbb{R}^{m \times n}$ *and* $b \in \mathbb{R}^m$. *The Lasserre hierarchy of level t,* $\mathrm{Las}_t(K)$, *is the set of vectors* $y \in \mathbb{R}^{\mathcal{I}_t}$, *that satisfy:*

$$y_0 = 1, \; M_t(y) \succeq 0, \; M_t^\ell(y) \succeq 0 \; \text{for every } \ell \in [m],$$

*where* $M_t(y) := (y_{I \cup J})_{|I|,|J| \le t}$ *is called the moment matrix and the matrices*

$$M_t^\ell(y) := \left( \sum_{i=1}^n A_{\ell i} y_{I \cup J \cup \{i\}} - b_\ell y_{I \cup J} \right)_{|I|,|J| \le t}$$

*are called* localizing matrices.

In general, a standard approach to tackle integer programs like (5.3) is to use *lift- and project methods*. We use Definition 3.2.1 to lift the solution space $K \subseteq \mathbb{R}^n$ of the original program to a higher dimensional space $\mathrm{Las}_t(K) \subseteq \mathbb{R}^{\mathcal{I}_t}$. Here we find an optimal solution $y \in \mathrm{Las}_t(K)$ and project it back to a set $L_t(K)$ satisfying $P \subseteq L_t(K) \subseteq K$. The sets $L_t(K)$ are as follows.

**Definition 3.2.2.** *Let* $y \in \mathrm{Las}_t(K)$ *and define the function* $\mathrm{proj} : \mathbb{R}^{\mathcal{I}_t} \to \mathbb{R}^{[n]}$ *by*

$$\mathrm{proj}(y) \mapsto (y_{\{1\}}, y_{\{2\}}, y_{\{3\}}, \dots, y_{\{n\}}).$$

*We denote its image on the original solution space* $x \in K$ *by*

$$\mathrm{Las}_t^{\mathrm{proj}}(K) := \{(y_{\{1\}}, \dots, y_{\{n\}}) : y \in \mathrm{Las}_t(K)\}.$$

Now, we have all definitions needed for (3.5), whereas a rigorous proof of (3.5) is due to the following Theorems 3.2.3 and 3.2.4. Because optimizing over $\mathrm{Las}_t(K)$ can often be executed in polynomial time in terms of the number and dimension of matrices $M_t(y)$ and $M_t^\ell(y)$ (see Section 2.1.1), one might wonder whether optimizing over $\mathrm{Las}_n(K)$ leads to an efficient algorithm to solve the NP-complete problem SET COVER. Thus it is important to note that the set of indices $\mathcal{I}_t = \{I \in \mathcal{P}([n]) : |I| \le t\}$ has exponential size $2^{O(t)}$ at step $\mathrm{Las}_t(K)$ in the hierarchy and consequently optimizing over $\mathrm{Las}_n(K)$ requires exponential time. However, for a variety of problems, such as INDEPENDENT SET, GRAPH COLORING (see [40]) or for finding error correcting codes (see [36]), even computing $\mathrm{Las}_2(K)$ provides significantly good bounds.

We begin to prove (3.5) by presenting a shortened version of a lemma in the survey of Rothvoß (see Lemma 1, [71]).

**Theorem 3.2.3.** *Let* $K = \{x \in \mathbb{R}^n : Ax \geq b\}$ *and* $y \in \text{Las}_t(K)$ *for* $t \geq 0$. *Then* $\text{Las}_t(K)$ *has the following properties:*

a) $K \cap \{0, 1\}^n \subseteq \text{Las}_t^{\text{proj}}(K)$

b) $\text{Las}_t^{\text{proj}}(K) \subseteq K$

c) $\text{Las}_0(K) \supseteq \text{Las}_1(K) \supseteq \ldots \supseteq \text{Las}_n(K)$

*Proof.* [71] We prove properties a) - c) as follows:

a) Let $x \in K \cap \{0, 1\}^n$ be a feasible integer solution. We prove that the vector $y \in \mathbb{R}^{I_t}$ defined by

$$y_I := \begin{cases} 1 & \text{if } I = \emptyset \\ \prod_{i \in I} x_i & \text{otherwise} \end{cases}$$

satisfies $y \in \text{Las}_t(K)$ for every $t \geq 0$. Thus we start checking the properties of Definition 3.2.1: The first property $y_\emptyset = 1$ is immediate from the definition above. For the moment matrix $M_t(y)$ we observe:

$$(M_t(y))_{I,J} = y_{I \cup J} = \prod_{i \in I \cup J} x_i = \prod_{i \in I} x_i \cdot \prod_{j \in J} x_j = y_I \cdot y_J = \left(yy^T\right)_{I,J},$$

where the third equality uses the fact that $x_i = x_i^2$ for every $x_i \in \{0, 1\}$. Thus $M_t(y)$ is actually a $2^t \times 2^t$ submatrix of the positive semidefinite matrix $yy^T$.
Finally we focus on the localizing matrices $M_t^\ell(y)$. Let us denote the $\ell$-th row in the system $Ax \geq b$ by $a^T x \geq \beta$ and conclude

$$\left(M_t^\ell(y)\right)_{I,J} = \sum_{i=1}^n a_i y_{I \cup J \cup \{i\}} - \beta y_{I \cup J} = \sum_{i=1}^n a_i y_{\{i\}} y_I y_J - \beta y_I y_J$$

$$= \left(\sum_{i=1}^n a_i x_i - \beta\right) y_I y_J = (a^T x - \beta) \cdot \left(yy^T\right)_{I,J}$$

Since $x$ is a feasible solution of $Ax \geq b$, we have that $a^T x - \beta \geq 0$ and thus $M_t^\ell(y) = (a^T x - \beta)(y_I y_J)_{|I|,|J| \leq t}$ is a scaled principal minor of the positive semidefinite matrix $yy^T$ implying positive semidefiniteness. Thus we verified the three properties and conclude $y \in \text{Las}_t(K)$ which closes the proof of a).

b) We check that $y' = (y_{\{1\}}, \ldots, y_{\{n\}})^T \in \text{Las}_t^{\text{proj}}$ satisfies every $\ell$-th constraint denoted by $a^T y' \geq \beta$. For this we consider the entry $M_t^\ell(y)_{0,0}$, which is nonnegative as a principal $1 \times 1$ submatrix of $M_t^\ell(y) \geq 0$, implying

$$0 \leq M_t^\ell(y)_{0,0} = \sum_{i=1}^n a_i y_{0 \cup \{i\}} - \beta y_0 = a^T y' - \beta.$$

c) We observe that the moment matrix $M_t(y)$ is a principal submatrix of $M_{t+1}(y)$ and that the same holds for $M_t^\ell(y)$ and $M_{t+1}^\ell(y)$ respectively. Thus for every solution $y \in \text{Las}_{t+1}(K)$ with $M_{t+1}(y) \succeq 0$ and $M_{t+1}^\ell(y) \succeq 0$ we also have $M_t(y) \succeq 0$ and $M_t^\ell(y) \succeq 0$ implying $y \in \text{Las}_t(K)$. The claim follows inductively.

$\square$

## 3.2.1   Convergence of the Lasserre hierarchy

The final goal of this paragraph is to prove *finite convergence*, i.e., the last equality in (3.5):

$$\text{Las}_n^{\text{proj}}(K) = \text{conv}(K \cap \{0,1\}^n).$$

Since $\text{Las}_n^{\text{proj}}(K)$ is convex, property a) in Theorem 3.2.3 implies $\text{Las}_n^{\text{proj}}(K) \supseteq \text{conv}(K \cap \{0,1\}^n)$. It remains to show that $\text{Las}_n^{\text{proj}}(K) \subseteq \text{conv}(K \cap \{0,1\}^n)$.

**Theorem 3.2.4.** *Given $y \in \text{Las}_t(K)$ and a subset of indices $S \subseteq [n]$ with $|S| \leq t$. Then,*

$$y \in \text{conv}\{z \in \text{Las}_{t-|S|}(K) : z_i \in \{0,1\} \text{ for every } i \in S\}.$$

Finite convergence follows by applying Theorem 3.2.4 to $S = [n]$: Then for $y \in \text{Las}_n(K)$ the theorem implies

$$y \in \text{conv}\{z \in \text{Las}_0(K) : z_i \in \{0,1\} \text{ for every } i \in [n]\},$$

or $\text{Las}_n(K) \subseteq \text{conv}\{z \in \text{Las}_0(K) : z_i \in \{0,1\} \text{ for every } i \in [n]\}$. If we project these sets by $\text{proj}: \mathbb{R}^{I_n} \to \mathbb{R}^{[n]}$ and recall that $\text{Las}_0^{\text{proj}}(K) \subseteq K$, we immediately obtain

$$\text{Las}_n^{\text{proj}}(K) \subseteq \text{conv}\{z \in \text{Las}_0^{\text{proj}}(K) : z_i \in \{0,1\} \text{ for every } i \in [n]\} \subseteq \text{conv}(K \cap \{0,1\}^n).$$

Thus after at most $n$ iterations the projected solution space of the Lasserre hierarchy has indeed converged to the integral hull of (3.4). In order to prove Theorem 3.2.4 we first prove Lemma 3.2.5, following the arguments in [71].

**Lemma 3.2.5.** *Let $t \geq 1$ and consider $y \in \text{Las}_t(K)$ with the property $0 < y_{\{i\}} < 1$ for an index $i \in [n]$. Then the vectors $z^{(0)}$ and $z^{(1)}$ defined by*

$$z_I^{(0)} := \frac{y_I - y_{I \cup \{i\}}}{1 - y_{\{i\}}}, \quad z_I^{(1)} := \frac{y_{I \cup \{i\}}}{y_{\{i\}}}$$

*satisfy the properties $z^{(0)}, z^{(1)} \in \text{Las}_{t-1}(K)$, $z_{\{i\}}^{(0)} = 0$, $z_{\{i\}}^{(1)} = 1$ and*

$$y = (1 - y_{\{i\}}) \cdot z^{(0)} + y_{\{i\}} \cdot z^{(1)}.$$

*Proof.* [71] Three properties hold immediately since $z_{\{i\}}^{(0)} = \frac{y_{\{i\}} - y_{\{i\}}}{1 - y_{\{i\}}} = 0$, $z_{\{i\}}^{(1)} = \frac{y_{\{i\}}}{y_{\{i\}}} = 1$ and

$$y_I = y_{I \cup \{i\}} + y_I - y_{I \cup \{i\}} = y_{\{i\}} \cdot \frac{y_{I \cup \{i\}}}{y_{\{i\}}} + (1 - y_{\{i\}}) \cdot \frac{y_I - y_{I \cup \{i\}}}{1 - y_{\{i\}}} = y_{\{i\}} \cdot z_I^{(1)} + (1 - y_{\{i\}}) \cdot z_I^{(0)}.$$

Thus it is left to show that $z^{(0)}, z^{(1)} \in \text{Las}_{t-1}(K)$. For this, we define the index sets $\mathcal{I}_-, \mathcal{I}_+ \subseteq \mathcal{I}_{t-1}$ by

$$\mathcal{I}_- := \{I : |I| < t, i \notin I\} \text{ and } \mathcal{I}_+ := \{I \cup \{i\} : |I| < t, i \notin I\}.$$

These sets are disjoint with $|\mathcal{I}_+| = |\mathcal{I}_-|$ and furthermore we have that the index set $\mathcal{I}_{t-1}$ of $\text{Las}_{t-1}(K)$ satisfies $\mathcal{I}_{t-1} \subseteq \mathcal{I}_- \cup \mathcal{I}_+$, since $\mathcal{I}_-$ includes all sets $I$ with $|I| \leq t - 1$ and $\{i\} \notin I$, whereas $\mathcal{I}_+$ includes all sets $I$ with $|I| \leq t - 1$ and $\{i\} \in I$. We consider the principal submatrices $M_{-i}$ and $M_{+i}$ of $M_t(y)$ indexed by $\mathcal{I}_-$ and $\mathcal{I}_+$. After rearranging the rows and columns of $M_t(y)$ we obtain

$$M_t(y) = \begin{array}{cc} & \begin{array}{ccc} \mathcal{I}_- & \mathcal{I}_+ & \end{array} \\ \begin{array}{c} \mathcal{I}_- \\ \mathcal{I}_+ \\ \end{array} & \left( \begin{array}{ccc} M_{-i} & M_{+i} & * \\ M_{+i} & M_{+i} & * \\ * & * & * \end{array} \right) \end{array}$$

$$= y_{\{i\}} \begin{pmatrix} \frac{1}{y_{\{i\}}} M_{+i} & \frac{1}{y_{\{i\}}} M_{+i} & * \\ \frac{1}{y_{\{i\}}} M_{+i} & \frac{1}{y_{\{i\}}} M_{+i} & * \\ * & * & * \end{pmatrix} + (1 - y_{\{i\}}) \begin{pmatrix} \frac{1}{1 - y_{\{i\}}}(M_{-i} - M_{+i}) & 0 & * \\ 0 & 0 & * \\ * & * & * \end{pmatrix}.$$

The entries of the first principal submatrix

$$M_1 := \begin{pmatrix} \frac{1}{y_{\{i\}}} M_{+i} & \frac{1}{y_{\{i\}}} M_{+i} \\ \frac{1}{y_{\{i\}}} M_{+i} & \frac{1}{y_{\{i\}}} M_{+i} \end{pmatrix}$$

are of the form $\frac{1}{y_{\{i\}}} \cdot y_{I \cup J \cup \{i\}}$ with $I, J \in \mathcal{I}_- \cup \mathcal{I}_+$. Thus the matrix $M_{t-1}(z^{(1)})$ with entries $z_{I \cup J}^{(1)} = \frac{1}{y_{\{i\}}} \cdot y_{I \cup J \cup \{i\}}$ for $I, J \in \mathcal{I}_{t-1} \subseteq \mathcal{I}_- \cup \mathcal{I}_+$ is a principal submatrix of $M_1$. Due to the fact that $M_{+i} \succeq 0$, the cloned and scaled copy $M_1$ is also positive semidefinite and thus every principal submatrix, in particular $M_{t-1}(z^{(1)})$, is positive semidefinite as well. The same idea applies for

$$M_2 := \begin{pmatrix} \frac{1}{1 - y_{\{i\}}}(M_{-i} - M_{+i}) & 0 \\ 0 & 0 \end{pmatrix}$$

having entries

$$(M_2)_{I,J} = \begin{cases} \frac{1}{1 - y_{\{i\}}} \cdot (y_{I \cup J} - y_{I \cup J \cup \{i\}}) & = z_{I \cup J}^{(0)} & \text{for } I, J \in \mathcal{I}_- \\ 0 & = z_{I \cup J}^{(0)} & \text{otherwise.} \end{cases}$$

We deduce similarly that $M_{t-1}(z^{(0)})$ is a principal submatrix of $M_2$ and show that $M_2$ is positive semidefinite:

$$\begin{pmatrix} x \\ x' \end{pmatrix}^T M_2 \begin{pmatrix} x \\ x' \end{pmatrix} = \begin{pmatrix} x \\ x' \end{pmatrix}^T \begin{pmatrix} (M_{-i} - M_{+i}) & 0 \\ 0 & 0 \end{pmatrix} \begin{pmatrix} x \\ x' \end{pmatrix} = x^T (M_{-i} - M_{+i}) x$$

$$= \begin{pmatrix} x \\ -x \end{pmatrix}^T \begin{pmatrix} M_{-i} & M_{+i} \\ M_{+i} & M_{+i} \end{pmatrix} \begin{pmatrix} x \\ -x \end{pmatrix} \geq 0,$$

where $x$ denotes the first $|\mathcal{I}_-|$ entries of $\begin{pmatrix} x \\ x' \end{pmatrix}$. Therefore $M_2 \geq 0$ and $M_{t-1}(z^{(0)}) \geq 0$ is a valid moment matrix.

Finally, for $z^{(0)}$ and $z^{(1)} \in \mathrm{Las}_{t-1}(K)$ it remains to show that the corresponding localizing matrices $M^\ell_{t-1}(z^{(0)})$ and $M^\ell_{t-1}(z^{(1)})$ are positive semidefinite. The proof uses the same arguments as above applied on every localizing matrix $M^\ell_t$ and its principal submatrices $M^\ell_{-i}$ and $M^\ell_{+i}$ indexed by $\mathcal{I}_-$ and $\mathcal{I}_+$ respectively:

$$M^\ell_t(y) = \begin{matrix} & \mathcal{I}_- & \mathcal{I}_+ & \\ \mathcal{I}_- \\ \mathcal{I}_+ \\ {} \end{matrix} \begin{pmatrix} M^\ell_{-i} & M^\ell_{+i} & * \\ M^\ell_{+i} & M^\ell_{+i} & * \\ * & * & * \end{pmatrix}$$

$$= y_{\{i\}} \begin{pmatrix} \frac{1}{y_{\{i\}}} M^\ell_{+i} & \frac{1}{y_{\{i\}}} M^\ell_{+i} & * \\ \frac{1}{y_{\{i\}}} M^\ell_{+i} & \frac{1}{y_{\{i\}}} M^\ell_{+i} & * \\ * & * & * \end{pmatrix} + (1 - y_{\{i\}}) \begin{pmatrix} \frac{1}{1-y_{\{i\}}}(M^\ell_{-i} - M^\ell_{+i}) & 0 & * \\ 0 & 0 & * \\ * & * & * \end{pmatrix}.$$

After having shown that $M_{t-1}(z^{(0)})$, $M^\ell_{t-1}(z^{(0)})$, $M_{t-1}(z^{(1)})$ and $M^\ell_{t-1}(z^{(1)}) \geq 0$, together with $z^{(0)}_0 = z^{(1)}_0 = 1$ by definition, we conclude $z^{(0)}, z^{(1)} \in \mathrm{Las}_{t-1}(K)$. $\qquad\Box$

For a fixed index $i \in [n]$, Lemma 3.2.5 shows that $y \in \mathrm{Las}_t(K)$ is contained in $\mathrm{conv}\{z \in \mathrm{Las}_{t-1}(K) : z_{\{i\}} \in \{0, 1\}\}$. Thus we can prove Theorem 3.2.4 by applying Lemma 3.2.5 iteratively for a fixed subset $S \subseteq [n]$, where for simplicity of notation we denote $S = \{1, \ldots, s\}$. The iterations then read:

$$\mathrm{Las}_t(K) \subseteq \mathrm{conv}\left\{\mathrm{Las}_{t-1}(K) \cap \left\{z \in \mathbb{R}^{\mathcal{I}_t} : z_{\{1\}} \in \{0, 1\}\right\}\right\}$$

$$\subseteq \mathrm{conv}\left\{\mathrm{Las}_{t-2}(K) \cap \left\{z \in \mathbb{R}^{\mathcal{I}_t} : z_{\{1\}}, z_{\{2\}} \in \{0, 1\}\right\}\right\}$$

$$\subseteq \vdots$$

$$\subseteq \mathrm{conv}\left\{\mathrm{Las}_{t-|S|}(K) \cap \left\{z \in \mathbb{R}^{\mathcal{I}_t} : z_{\{1\}}, \ldots, z_{\{s\}} \in \{0, 1\}\right\}\right\}.$$

### 3.2.2    Applying the Lasserre hierarchy to SET COVER

In Section 3.1.2 it was shown that a simple greedy heuristic already yields a fairly good approximation ratio of SET COVER, i.e., it finds a solution with objective value $c^T x \leq SC \cdot H(M)$, where $M$ denotes the size of the largest subset. This leads to the observation that the greedy algorithm provides better approximations, the smaller the largest set is. In this section, we will exploit this fact by utilizing the Lasserre hierarchy to a priori put "good sets" into the cover and let the greedy heuristic act only on the remaining small sets. The resulting algorithm is due to Chlamtác, Friggstad and Georgiou [12] and achieves an approximation ratio of $(1 - \varepsilon)\ln(m) + o(1)$, where $0 < \varepsilon < 1$. This ratio is better than the one of the greedy heuristic but needs subexponential runtime instead of polynomial runtime. Hereafter, we again follow the presentation in Rothvoß' [71] lecture notes and assume all entries of $c, A, b$ to be rational.

We consider the linear relaxation of $SC$ and instead of applying Lasserre's hierarchy directly, we add a constraint $c^T x \leq q$ with a predetermined (see below) constant $q \leq SC$ to obtain

$$K_q := \left\{ x \in \mathbb{R}^n : \sum_{i:k \in S_i} x_i \geq 1 \text{ for every } k \in [m], c^T x \leq q, \ 0 \leq x_i \leq 1 \text{ for every } i \in [n] \right\}.$$

For every fixed $0 < \varepsilon < 1$ such that $m^\varepsilon \in \mathbb{N}$, the constant $q$ is determined via binary search: Without loss of generality we assume $c \in \mathbb{N}^n$, if otherwise we rescale the constraint $c^T x \leq q$ with the product of the denominators of $c$. The goal is to find the smallest $q$ for which there is still a feasible solution $y \in \mathrm{Las}_{m^\varepsilon}(K_q)$, where $m$ is the number of covering constraints. We begin by setting $q = 2^0$ and check whether $\mathrm{Las}_{m^\varepsilon}(K_q) \neq \emptyset$. If $\mathrm{Las}_{m^\varepsilon}(K_q) \neq \emptyset$, we stop. If not, we iterate the whole procedure with $q = 2^k$ and $k = 1, 2, \ldots$. Once we have found a $q = 2^k$ satisfying $\mathrm{Las}_{m^\varepsilon}(K_q) \neq \emptyset$, we continue searching for a smaller $q \in [2^{k-1}, 2^k]$, for which this also holds, by performing a binary search. Thus we end up with the smallest $q \in \mathbb{N}$ for which $\mathrm{Las}_{m^\varepsilon}(K_q) \neq \emptyset$ holds.

With the help of the computed $q$ and its corresponding relaxation $K_q$ we can now approximate $SC$ with the following Algorithm 2. To show that this algorithm requires subexponential time to find a $(1 - \varepsilon)\ln(m) + o(1)$ approximation of SET COVER we present the proof of Rothvoß [71]. It is important to note that computing a feasible $y \in \mathrm{Las}_t(K_q)$ is computable in polynomial time in terms of the dimension of the block diagonal matrix

$$\begin{pmatrix} M_t(y) & 0 & \cdots & 0 \\ 0 & M_t^1(y) & \ddots & \vdots \\ \vdots & \ddots & \ddots & 0 \\ 0 & \cdots & 0 & M_t^{m+1}(y) \end{pmatrix} \in S^{(m+2) \cdot O(t \cdot n^t)}.$$

This fact boils down to finding a solution for the weak optimization problem, defined in Chapter 2, on the bounded convex body $\mathrm{Las}_{m^\varepsilon}(K_q)$. The weak optimization problem can be solved in $n^{O(m^\varepsilon)}$ time due to Example 2.1.7.

---

**Algorithm 2** Algorithm using the Lasserre hierarchy for SET COVER

---

**Require:** Ground set $[m]$, subsets $S_1, \ldots, S_n$, costs $c_1, \ldots, c_n$
**Ensure:** Cover $\mathcal{J}$
1: Covered elements $C \leftarrow \emptyset$
2: $\mathcal{J} \leftarrow \emptyset$
3: Find a $q$ for which $\mathrm{Las}_{m^\varepsilon}(K_q) \neq \emptyset$ (via binary search)
4: Compute $y^{(0)} \in \mathrm{Las}_{m^\varepsilon}(K_q)$
5: **for** $r \in 1, \ldots, m^\varepsilon$ **do**
6:     Choose $j \in [n] \setminus \mathcal{J}$ with $y_{\{j\}}^{(r-1)} > 0$ and $|S_j \setminus C| \geq |S_i \setminus C|$ for every $i \in [n]$
7:     $\mathcal{J} \leftarrow \mathcal{J} \cup \{j\}$
8:     $C \leftarrow C \cup S_j$
9:     $y_I^{(r)} \leftarrow \dfrac{y_{I \cup \{j\}}^{(r-1)}}{y_{\{j\}}^{(r-1)}}$ (see Lemma 3.2.5)
10:    $S_i \leftarrow S_i \setminus S_j$ for every $i \in [n] \setminus \mathcal{J}$
11: **end for**
12: Find cover $\mathcal{J}'$ for the remaining elements, using Algorithm 1
13: $\mathcal{J} \leftarrow \mathcal{J} \cup \mathcal{J}'$

---

**Theorem 3.2.6.** *For a fixed $\varepsilon$ with $0 < \varepsilon < 1$, Algorithm 2 provides an approximation for* SET COVER *with cost at most $((1 - \varepsilon)\ln(m) + o(1)) \cdot SC$ in time $n^{O(m^\varepsilon)}$.*

*Proof.*

1) *Time:*
Since computing a feasible $y \in \mathrm{Las}_{m^\varepsilon}(K_q)$ is possible in $n^{O(m^\varepsilon)}$ time and the binary search needs $O(\log(SC))$ iterations only depending on the optimal value $SC \leq n$, we need $n^{O(m^\varepsilon)}$ time to find $q$. The loop always runs in polynomial time, as does the greedy algorithm. Thus the resulting time of the entire algorithm lies in $n^{O(m^\varepsilon)}$.

2) *Quality of the approximation:*
After computing $q$ with the property $\mathrm{Las}_{m^\varepsilon}(K_{q-1}) = \emptyset$ via binary search, we conclude via Theorem 3.2.3

$$\emptyset = \mathrm{Las}_{m^\varepsilon}(K_{q-1}) \supseteq \mathrm{Las}_{m^\varepsilon+1}(K_{q-1}) \supseteq \cdots \supseteq \mathrm{Las}_n(K_{q-1}) = \mathrm{conv}\{K_{q-1} \cap \{0,1\}^n\}$$

and thus $\mathrm{conv}\{K_{q-1} \cap \{0,1\}^n\} = \emptyset$. This implies $q - 1 < SC$ and further $q \leq SC$ due to integrality of rescaled $c^T x$ and $q$.
We further observe that for every index $j \in \mathcal{J} = \{j_1, \ldots, j_{m^\varepsilon}\}$, where $j_r$ denotes the index picked in iteration $r$, we have that

$$y_{\{j\}}^{(r)} = \frac{y_{\{j\} \cup \{j_r\}}^{(r-1)}}{y_{\{j_r\}}^{(r-1)}} = \frac{y_{\{j, j_r\} \cup \{j_{r-1}\}}^{(r-2)}}{y_{\{j_r\}}^{(r-2)}} \cdot \frac{y_{\{j_{r-1}\}}^{(r-2)}}{y_{\{j_r\} \cup \{j_{r-1}\}}^{(r-2)}} = \frac{y_{\{j\} \cup \{j_r, j_{r-1}\}}^{(r-2)}}{y_{\{j_r, j_{r-1}\}}^{(r-2)}}$$

and conclude inductively $y_{\{j\}}^{(m^\varepsilon)} = \frac{y_{\mathcal{J}}^{(0)}}{y_{\mathcal{J}}^{(0)}} = 1$. Together with the fact that inductively applying Lemma 3.2.5 shows that $y^{(m^\varepsilon)} \in \text{Las}_0(K_q)$, i.e., $y' := \text{proj}(y^{(m^\varepsilon)}) \in \text{Las}_0^{\text{proj}}(K_q) = K_q$, we conclude that

$$c^T y' = \sum_{j \in \mathcal{J}} c_j + \sum_{j \notin \mathcal{J}} c_j y'_j \le q \le SC.$$

To cover the remaining elements, $X' := [m] \setminus C$, we apply the greedy Algorithm 1. Since $y' \in K_q$ defines a fractional cover, we can find a proper cover by restricting to the sets $S' = \{S_i \setminus C : y'_i > 0\}$ pickable by Algorithm 2 in a hypothetical $(m^\varepsilon + 1)$-th iteration. But for every set $S_j$ with $j \in \mathcal{J}$ we know from line 6 in Algorithm 2 that $|S_j| \ge |S_i \setminus C|$ for $S_i \setminus C \in S'$. This implies that $|S_i| \le m^{1-\varepsilon}$ since otherwise we would already have covered $\sum_{j \in \mathcal{J}} |S_j| \ge m^\varepsilon \cdot m^{1-\varepsilon} = m$ elements.

Finally, we apply the greedy heuristic achieving a cover for the remaining sets of costs at most $(\ln(m^{1-\varepsilon}) + 1) \sum_{j \notin \mathcal{J}} c_j y'_j$ and we end up with a cover of costs

$$\sum_{j \in \mathcal{J}} c_j + (\ln(m^{1-\varepsilon}) + 1) \sum_{j \notin \mathcal{J}} c_j y'_j \le c^T y' + \ln(m^{1-\varepsilon}) SC \le ((1-\varepsilon)\ln(m) + 1) SC.$$

$\square$

## 3.3   Exploiting symmetries in SET COVER

In this section we are interested in symmetric instances of SET COVER; that is, we assume $n = m$, $i \in S_j \Leftrightarrow j \in S_i$ and the existence of a group $\Gamma$ with a transitive action on $[n]$ such that

$$\{S_1, \dots, S_n\} = \{\gamma S_1 : \gamma \in \Gamma\}.$$

In particular, transitivity implies $|S_i| = |S_1|$ for every $i \in [n]$.

These instances are essentially the finite versions of the upcoming covering problems in Chapter 5 and the techniques we illustrate here turn out to hold in the setting of compact metric spaces as well. Symmetry reduction is not part of Rothvoß' lecture notes [71]. However, if applicable, this technique is a powerful tool to reduce the number of constraints in the Lasserre hierarchy significantly.

The integer linear program formulation for such symmetric problems reads

$$SSC = \min\left\{ \sum_{i=1}^n x_i : x \in \{0,1\}^n, \sum_{i \in \gamma S_1} x_i \ge 1 \text{ for every } \gamma \in \Gamma \right\} \tag{3.6}$$

and the corresponding linear programming relaxation

$$SSC\mathcal{LP} := \min\left\{ \sum_{i=1}^n x_i : x \in [0,1]^n, \sum_{i \in \gamma S_1} x_i \ge 1 \text{ for every } \gamma \in \Gamma \right\} \tag{3.7}$$

of $SSC$ gives a trivial bound: The vector $x$ defined by $x_i = 1/|S_1|$ for all $i \in [n]$ is feasible and has objective value $n/|S_1|$, so $SSC\mathcal{LP} \le \frac{n}{|S_1|}$. For the inequality $SSC\mathcal{LP} \ge \frac{n}{|S_1|}$ we observe that for feasible $x$, the vector $\bar{x}$, defined by

$$\bar{x}_i = \frac{1}{|\Gamma|} \sum_{\gamma \in \Gamma} x_{\gamma i}, \quad \text{for} \quad i \in [n],$$

is also feasible and has the same objective value. Since the action of $\Gamma$ on $[n]$ is transitive, we have $\bar{x}_i = \bar{x}_j$ for all $i$ and $j$, and hence $\bar{x}_i \ge 1/|S_1|$ for all $i$, which implies $SSC\mathcal{LP} \ge \frac{n}{|S_1|}$. This resembles the volume bound (1.1).

We apply the Lasserre hierarchy by defining for the vectors $y \in \mathbb{R}^{I_t}$ the moment matrix $M(y)$ by $(M(y))_{J,J'} := y_{J \cup J'}$ and its submatrix $M_t(y)$ by restricting the cardinality of the sets $J, J'$ to $|J|, |J'| \le t$. The localizing matrices $M_t^\gamma(y)$ are defined by

$$\left( M_t^\gamma(y) \right)_{J,J'} := \left( \sum_{i \in \gamma S_1} y_{J \cup J' \cup \{i\}} - y_{J \cup J'} \right)_{|J|,|J'| \le t}.$$

The Lasserre hierarchy over $K_\Gamma = \{x \in [0,1]^n : \sum_{i \in \gamma S_1} x_i \ge 1 \text{ for every } \gamma \in \Gamma\}$ now equals

$$\text{Las}_t(K_\Gamma) = \min \left\{ \sum_{i=1}^n y_{\{i\}} : y \in \mathbb{R}^{I_{2t}}, \ y_\emptyset = 1, \ M_t(y) \ge 0, \ M_t^\gamma(y) \ge 0 \text{ for } \gamma \in \Gamma \right\}, \quad (3.8)$$

where $I_{2t} = \{I \subseteq [n] : |I| \le 2t\}$. As we have seen in the sections before it is notoriously costly to compute higher steps in the hierarchy. A convenient tool to simplify such computations is to apply *symmetry reduction*, i.e., consider the symmetry inherent in the program and restrict the program to solutions invariant under this symmetry.

For a feasible solution $y \in \mathbb{R}^{I_{2t}}$ we show that the vector $\bar{y} \in \mathbb{R}^{I_{2t}}$ defined by $\bar{y}_J := \frac{1}{|\Gamma|} \sum_{\gamma \in \Gamma} y_{\gamma J}$, where $\gamma J = \{\gamma j : j \in J\}$, is also a feasible solution for (3.8) with the same objective value $\sum_{i=1}^n \bar{y}_{\{i\}} = \sum_{i=1}^n y_{\{i\}}$. For this we observe that $\gamma(A \cup B) = \gamma A \cup \gamma B$ for any two finite sets $A, B$ and

$$\sum_{i=1}^n \bar{y}_{\{i\}} = \frac{1}{|\Gamma|} \sum_{\gamma \in \Gamma} \sum_{i=1}^n y_{\{\gamma i\}} = \sum_{i=1}^n y_{\{i\}},$$

$$\bar{y}_\emptyset = y_\emptyset = 1,$$

$$M_t(\bar{y})_{J,J'} = \bar{y}_{J \cup J'} = \frac{1}{|\Gamma|} \sum_{\gamma \in \Gamma} y_{\gamma J \cup \gamma J'} = \frac{1}{|\Gamma|} \sum_{\gamma \in \Gamma} M_t(y)_{\gamma J, \gamma J'}.$$

For the localizing matrices indexed by $\gamma' \in \Gamma$, we have

$$M_t^{\gamma'}(\bar{y})_{J,J'} = \frac{1}{|\Gamma|} \sum_{\gamma \in \Gamma} \left( \sum_{i \in \gamma' S_1} y_{\gamma J \cup \gamma J' \cup \{\gamma i\}} - y_{\gamma J \cup \gamma J'} \right) = \frac{1}{|\Gamma|} \sum_{\gamma \in \Gamma} M_t^{\gamma \circ \gamma'}(y)_{\gamma J, \gamma J'}.$$

We further know that simultaneous row and column permutations do not change the eigen-values of a matrix and thus the matrices $M_t(\bar{y})$ and $M_t^\gamma(\bar{y})$ remain positive semidefinite. Thus we can restrict the program (3.8) to the $\Gamma$-invariant vectors $\bar{y} \in \mathbb{R}^{I_{2t}}$ without changing the optimal value and conclude

$$\text{Las}_t(K_\Gamma) = \min\left\{\sum_{i=1}^n y_{\{i\}} : y \in \mathbb{R}^{I_{2t}/\Gamma}, \ y_\emptyset = 1, \ M_t(y) \succeq 0, \ M_t^\gamma(y) \succeq 0 \text{ for } \gamma \in \Gamma\right\},$$

where $I_{2t}/\Gamma$ denotes the quotient space defined by the equivalency relation $I \sim J$ whenever there is a $\gamma \in \Gamma$ such that $I = \gamma J$. For our final simplification we first observe that for $\Gamma$-invariant vectors $y$, i.e., $y_J = y_{\gamma J}$, we also have

$$M_t^\gamma(y)_{J,J'} = \sum_{i \in \gamma S_1} y_{J \cup J' \cup \{i\}} - y_{J \cup J'} = \sum_{i \in S_1} y_{J \cup J' \cup \{\gamma i\}} - y_{J \cup J'}$$

$$= \sum_{i \in S_1} y_{\gamma^{-1} J \cup \gamma^{-1} J' \cup \{i\}} - y_{\gamma^{-1} J \cup \gamma^{-1} J'} = M_t^{id}(y)_{\gamma^{-1} J, \gamma^{-1} J'}.$$

Consequently for $\Gamma$-invariant vectors $y$ we have that

$$M_t^{id}(y) \succeq 0 \Leftrightarrow M_t^\gamma(y) \succeq 0 \text{ for every } \gamma \in \Gamma$$

leading to the following

$$\text{Las}_t(K_\Gamma) = \min\left\{\sum_{i=1}^n y_{\{i\}} : y \in \mathbb{R}^{I_{2t}/\Gamma}, \ y_\emptyset = 1, \ M_t(y) \succeq 0, \ M_t^{id}(y) \succeq 0\right\}. \tag{3.9}$$

This reformulation of the Lasserre hierarchy is computationally faster due to the fact that we have reduced the dimension of the matrices from $(n+1) \cdot O(t \cdot n^t)$ to $2 \cdot O(t \cdot n^t)$ and the number of variables from $|I_{2t}|$ to $|I_{2t}/\Gamma|$. In the remainder of this thesis we study geometric covering problems of a structure similar to SET COVER. A possible way of tackling these problems would be to sample the underlying geometric structure and work with the corresponding SET COVER problem as was done by Naszódi [60] to obtain upper bounds. However, the main idea in Chapter 5 is to work with an infinite analogue of the symmetrized version of Lasserre's hierarchy to keep the geometric properties and achieve lower bounds.

CHAPTER FOUR

# Upper bounds

This chapter is based on the publication *Covering compact metric spaces greedily, by J. H. Rolfes and F. Vallentin,* Acta Math. Hungar., 155(1):130–140, 2018."

A general greedy approach to construct coverings of compact metric spaces by metric balls is given and analyzed. The analysis is a continuous version of Chvátal's analysis of the greedy algorithm for the weighted set cover problem illustrated in Section 3.1.2. The approach is demonstrated in an exemplary manner to construct efficient coverings of the $n$-dimensional sphere and $n$-dimensional Euclidean space to give short and transparent proofs of several best known bounds obtained from deterministic constructions in the literature on sphere coverings.

## 4.1 Introduction

Let $X$ be a compact metric space having metric $d$. Given a scalar $r \in \mathbb{R}_{\geq 0}$ we define the *closed ball* of radius $r$ around center $x \in X$ by

$$B(x, r) = \{y \in X : d(x, y) \leq r\}.$$

The *covering number* of the space $X$ and a positive number $r$ is

$$\mathcal{N}(X, r) = \min \left\{ |Y| : Y \subseteq X, \bigcup_{y \in Y} B(y, r) = X \right\},$$

i.e., it is the smallest number of balls with radius $r$ one needs to cover $X$. Determining the covering number is a fundamental problem in metric geometry (see for example the classical book by Rogers [68]) with many applications: compressive sensing [34], approximation theory and machine learning [17] — to name a few.

In this chapter we are concerned with compact metric spaces which carry a probability measure $\omega$; a Borel measure normalized by $\omega(X) = 1$. We will assume that this probability measure is non degenerate and behaves homogeneously on balls, i.e., it satisfies the following two conditions:

(a) $\omega(B(x, s)) = \omega(B(y, s))$ for all $x, y \in X$, and for all $s \geq 0$,

(b) $\omega(B(x, s)) < \omega(B(x, t))$ whenever $B(x, s)$ is strictly contained in $B(x, t)$.

By (a) the measure of a ball does only depend on the radius $s$ and not on the center $x$, so we simply denote $\omega(B(x, s))$ by $\omega_s$ throughout the paper.

**Theorem 4.1.1.** *Let $(X, d)$ be a compact metric space with probability measure $\omega$ satisfying conditions (a) and (b). Then for every $\varepsilon$ with $r/2 > \varepsilon > 0$ the covering number satisfies*

$$\frac{1}{\omega_r} \leq \mathcal{N}(X, r) \leq \frac{1}{\omega_{r-\varepsilon}} \left( \ln\left( \frac{\omega_{r-\varepsilon}}{\omega_\varepsilon} \right) + 1 \right).$$

The lower bound is obvious (using the $\sigma$-subadditivity of $\omega$). We give a proof for the upper bound in Section 4.2. Our proof is based on a greedy approach to covering. We iteratively choose balls which cover the maximum measure of yet uncovered space.

We recall that this greedy algorithm has been analyzed in the finite setting of the SET COVER problem which is a fundamental problem in combinatorial optimization. The SET COVER problem is defined as follows. Given a collection $S_1, \ldots, S_n$ of the ground set $\{1, \ldots, m\}$ and given costs $c_1, \ldots, c_n$ the task is find a set of indices $I \subseteq \{1, \ldots, n\}$ such that $\bigcup_{i \in I} S_i = \{1, \ldots, m\}$ and $\sum_{i \in I} c_i$ is as small as possible.

Computationally, the SET COVER problem is difficult; Dinur and Steurer [27] showed that for every $\varepsilon > 0$ it is NP-hard to find an approximation to the SET COVER problem within a factor of $(1 - \varepsilon) \ln m$. On the other hand, Chvátal [13] (previously, Johnson [47], Stein [78] and Lovász [57] proved similar results for the case of uniform costs $c_1 = \ldots = c_n = 1$) showed that the greedy algorithm gives an $(\ln m + 1)$-approximation for the SET COVER problem. More specifically, Chvátal showed that the natural linear programming relaxation of SET COVER

$$\min \sum_{i=1}^{n} c_i x_i :$$

$$x_1, \ldots, x_n \geq 0,$$

$$\sum_{i: j \in S_i} x_i \geq 1 \text{ for all } j = 1, \ldots, m$$

is at most a factor of $H(M) = \sum_{j=1}^{M} \frac{1}{j} \leq \ln M + 1$, with $M = \max_i |S_i|$, away from an optimal solution of SET COVER. In Section 3.1.2 we provided his proof of this bound by exhibiting an appropriate feasible solution of the dual of the linear programming relaxation. The greedy algorithm was used to construct this feasible solution. We want to stress that although the techniques in Chapter 3 are designed for a finite ground set, they turn out to work in a similar way for compact metric spaces. For this it is important to note, that the analogue of a set of one element in the ground set $[m]$ is the (small) ball $B(x, \varepsilon)$, since we need every set in our cover to have positive volume.

In Section 4.2 we transfer Chvátal's argument from the finite SET COVER setting to the setting of compact metric spaces. Function $g$ appearing there features the feasible solution of the dual linear program. This will provide a proof of Theorem 4.1.1. In Section 4.3 we apply Theorem 4.1.1 to three concrete geometric settings and we retrieve some of the best known asymptotic results, unifying many results on sphere coverings.

We think that the NP-hardness of getting $(1 - \varepsilon) \ln m$-approximations for the SET COVER problem is a natural barrier for getting better asymptotic results for geometric covering problems. This might serve as an explanation why progress for example on the sphere covering problem has been very slow since the initial work of Rogers [68].

We are not the first observing the strong relation between geometric covering problems and SET COVER[1]. In recent papers, Artstein-Avidan and Raz [4], Artstein-Avidan and Slomka [5] and especially Naszódi [60] used the results of Lovász [57] to unify old results and prove new results on geometric coverings. However, they apply the results from SET COVER directly after choosing a finite $\varepsilon$-net. Since we consider an infinite analogue of SET COVER we do not need to use an $\varepsilon$-net and by this we sometimes get slightly better constants and more importantly we think that the analysis becomes rather beautiful.

## 4.2 Proof of Theorem 4.1.1

We shall prove that the following greedy algorithm (Algorithm 3) will provide a covering of $X$ with at most

$$\frac{1}{\omega_{r-\varepsilon}} \left( \ln\left( \frac{\omega_{r-\varepsilon}}{\omega_\varepsilon} \right) + 1 \right)$$

many balls of radius $r$.

---

**Algorithm 3** Greedy algorithm

---

1: $i \leftarrow 0$
2: $S^i_x = B(x, r - \varepsilon)$ for all $x \in X$
3: **while** $\bigcup_{j=1}^i B(y^j, r) \neq X$ **do**
4:     $i \leftarrow i + 1$
5:     Choose $y \in X$ with $\omega(S^{i-1}_y) \geq \omega(S^{i-1}_x)$ for all $x \in X$
6:     $y^i \leftarrow y$
7:     $S^i_x \leftarrow S^{i-1}_x \setminus S^{i-1}_y$ for all $x \in X$
8: **end while**

---

[1]In fact, we realized this only after we, in an attempt to understand geometric covering problems from an optimization point of view, wrote down the main body of this paper.

We split the proof into three lemmas where the following identity will become important:

$$S_x^{i-1} = B(x, r - \varepsilon) \setminus \bigcup_{j=1}^{i-1} B(y^j, r - \varepsilon). \tag{4.1}$$

The first lemma states that the step of the algorithm when we want to choose $y \in X$, with $\omega(S_y^{i-1}) \geq \omega(S_x^{i-1})$ for all $x \in X$, is indeed well-defined.

**Lemma 4.2.1.** *In every iteration $i$ the supremum* $\sup\{\omega(S_x^{i-1}) : x \in X\}$ *is attained.*

*Proof.* We shall show that the function $f_i \colon X \to \mathbb{R}$, $f_i(x) = \omega(S_x^{i-1})$ is continuous for every iteration $i$. This implies that $f_i$ attains its maximum since $X$ is compact.
For $x, y \in X$ we have

$$
\begin{aligned}
|f_i(x) - f_i(y)| &= |\omega(S_x^{i-1}) - \omega(S_y^{i-1})| \\
&= |\omega(S_x^{i-1} \setminus S_y^{i-1}) + \omega(S_x^{i-1} \cap S_y^{i-1}) \\
&\quad - (\omega(S_y^{i-1} \setminus S_x^{i-1}) + \omega(S_y^{i-1} \cap S_x^{i-1}))| \\
&= |\omega(S_x^{i-1} \setminus S_y^{i-1}) - \omega(S_y^{i-1} \setminus S_x^{i-1})| \\
&\leq \max\{\omega(S_x^{i-1} \setminus S_y^{i-1}), \omega(S_y^{i-1} \setminus S_x^{i-1})\}.
\end{aligned}
$$

Without loss of generality, the maximum is attained at $\omega(S_x^{i-1} \setminus S_y^{i-1})$. Then by (4.1) we see

$$S_x^{i-1} \setminus S_y^{i-1} \subseteq B(x, r - \varepsilon) \setminus B(y, r - \varepsilon).$$

By the triangle inequality

$$B(x, r - \varepsilon) \setminus B(y, r - \varepsilon) \subseteq B(y, r - \varepsilon + d(x, y)) \setminus B(y, r - \varepsilon).$$

Now consider the indicator function $\mathbb{1}_{B(y, r - \varepsilon + d(x,y)) \setminus B(y, r - \varepsilon)}$. When $y$ tends to $x$, then we have a monotonously decreasing sequence of measurable functions tending to 0. By applying the theorem of monotone convergence we obtain that the integral

$$\int \mathbb{1}_{B(y, r - \varepsilon + d(x,y)) \setminus B(y, r - \varepsilon)}(z) \, d\omega(z)$$

tends to 0 as well. Hence, $f_i(y)$ tends to $f_i(x)$. □

The second lemma states that the algorithm terminates after finitely many iterations.

**Lemma 4.2.2.** *Algorithm 3 terminates after at most $\omega_\varepsilon^{-1}$ iterations and returns a covering.*

*Proof.* Consider the $i$-th iteration of the algorithm and suppose there exists $x \in X$ with $x \notin \bigcup_{j=1}^{i-1} B(y^j, r)$. From the triangle inequality it follows that

$$B(x, \varepsilon) \cap B(y^j, r - \varepsilon) = \emptyset.$$

Choose $y \in X$ with $\omega(S_y^{i-1}) \geq \omega(S_x^{i-1})$ for every $x \in X$. Hence we have

$$\omega(S_y^{i-1}) \geq \omega(S_x^{i-1}) \geq \omega(B(x, \varepsilon)) = \omega_\varepsilon > 0$$

and thus

$$1 = \omega(X) \geq \sum_{j=1}^{i} \omega(S_{y^j}^{j-1}) \geq i \cdot \omega_\varepsilon,$$

where the first inequality follows because the sets $S_{y^j}^{j-1}$, with $j = 1, \ldots, i$, are pairwise disjoint. So after at most $\omega_\varepsilon^{-1}$ iterations, the algorithm terminates with a covering.    □

The third lemma gives the desired upper bound for the covering number.

**Lemma 4.2.3.** *Algorithm 3 terminates after at most*

$$\frac{1}{\omega_{r-\varepsilon}} \left( \ln \left( \frac{\omega_{r-\varepsilon}}{\omega_\varepsilon} \right) + 1 \right)$$

*iterations. In particular, this number gives an upper bound for the covering number* $N(X, r)$.

*Proof.* Let $Y \subseteq X$ denote the covering produced by Algorithm 3 after $|Y|$ iterations. We shall prove

$$\ln \left( \frac{\omega_{r-\varepsilon}}{\omega_\varepsilon} \right) + 1 \geq |Y| \cdot \omega_{r-\varepsilon}. \tag{4.2}$$

For this we define the symmetric kernel $K \colon X \times X \to \mathbb{R}$ by

$$K(x, y) = \begin{cases} 1, & \text{if } y \in B(x, r - \varepsilon) \\ 0, & \text{otherwise.} \end{cases}$$

For every $x \in X$ the following equality

$$\int K(x, y) \, d\omega(y) = \omega_{r-\varepsilon}$$

holds, where the integral exists due to Lebesgue's dominated convergence theorem, as we have a bounded function on a compact domain.

We will exhibit an integrable function $g \colon X \to \mathbb{R}$ satisfying

$$\int K(x, y) g(x) \, d\omega(x) \leq \ln \left( \frac{\omega_{r-\varepsilon}}{\omega_\varepsilon} \right) + 1 \tag{4.3}$$

for all $y \in X$ and satisfying

$$\int g(x) \, d\omega(x) = |Y|. \tag{4.4}$$

Combining (4.3) and (4.4), we get

$$\ln\left(\frac{\omega_{r-\varepsilon}}{\omega_\varepsilon}\right) + 1 \geq \int \int K(x,y)g(x)\,d\omega(x)d\omega(y)$$

$$= \int g(x) \int K(x,y)\,d\omega(y)d\omega(x)$$

$$= \int g(x)\omega_{r-\varepsilon}\,d\omega(x)$$

$$= |Y| \cdot \omega_{r-\varepsilon}$$

and we have proven (4.2). Now we only have to exhibit the function $g$. For brevity, we denote $\omega_y^{i-1} = \omega(S_y^{i-1})$. We define $g$ as follows:

$$g(x) = \begin{cases} (\omega_{y^i}^{i-1})^{-1}, & \text{if } x \in S_{y^i}^{i-1}, \\ 0, & \text{otherwise,} \end{cases}$$

which is a valid definition since the sets $S_{y^i}^{i-1}$ are pairwise disjoint. Also observe that $g$ is an integrable function on the compact set $X$. From this definition of $g$ we immediately get (4.4):

$$\int g(x)\,d\omega(x) = \sum_{i=1}^{|Y|} \omega_{y^i}^{i-1}(\omega_{y^i}^{i-1})^{-1} = |Y|.$$

To prove (4.3) we fix $y \in X$. We observe the equality

$$B(y, r-\varepsilon) \cap S_{y^i}^{i-1} = S_y^{i-1} \setminus S_y^i,$$

which describes which part of $B(y, r-\varepsilon)$ is cut away in iteration $i$. Then,

$$\int K(x,y)g(x)\,d\omega(x) = \sum_{i=1}^{|Y|} \int K(x,y)\mathbb{1}_{S_{y^i}^{i-1}}(x)(\omega_{y^i}^{i-1})^{-1}\,d\omega(x)$$

$$= \sum_{i=1}^{|Y|} \int \mathbb{1}_{S_y^{i-1}\setminus S_y^i}(x)(\omega_{y^i}^{i-1})^{-1}\,d\omega(x)$$

$$= \sum_{i=1}^{|Y|} (\omega_y^{i-1} - \omega_y^i)(\omega_{y^i}^{i-1})^{-1}.$$

For $y \in X$ consider the last iteration $b$ such that

$$\omega_{r-\varepsilon} = \omega(B(y, r-\varepsilon)) = \omega_y^1 \geq \ldots \geq \omega_y^b \geq \omega(B(y,\varepsilon)) = \omega_\varepsilon \qquad (4.5)$$

holds (here we used $r/2 > \varepsilon$). Note that $b < |Y|$. Note also that $\omega_y^{i-1} \leq \omega_{y^i}^{i-1}$ holds. We split the sum above into two parts:

$$\sum_{i=1}^{|Y|}(\omega_y^{i-1} - \omega_y^i)(\omega_{y^i}^{i-1})^{-1} = \sum_{i=1}^{b}(\omega_y^{i-1} - \omega_y^i)(\omega_{y^i}^{i-1})^{-1} + \sum_{i=b+1}^{|Y|}(\omega_y^{i-1} - \omega_y^i)(\omega_{y^i}^{i-1})^{-1}$$

$$\leq \sum_{i=1}^{b}(\omega_y^{i-1} - \omega_y^i)(\omega_y^{i-1})^{-1} + (\omega_y^b - \omega_y^{b+1})(\omega_y^b)^{-1}$$

$$+ \sum_{i=b+2}^{|Y|}(\omega_y^{i-1} - \omega_y^i)\omega_\varepsilon^{-1}$$

$$\leq \left(\sum_{i=1}^{b}(\omega_y^{i-1} - \omega_y^i)(\omega_y^{i-1})^{-1} + \frac{\omega_y^b - \omega_\varepsilon}{\omega_y^b}\right)$$

$$+ \left(\frac{\omega_\varepsilon - \omega_y^{b+1}}{\omega_\varepsilon} + \frac{\omega_y^{b+1} - \omega_y^{|Y|}}{\omega_\varepsilon}\right).$$

The first sum is a lower Riemann sum of the function $x \mapsto \frac{1}{x}$ in the interval $[\omega_\varepsilon, \omega_{r-\varepsilon}]$ and thus we have $\ln\left(\frac{\omega_{r-\varepsilon}}{\omega_\varepsilon}\right)$ as an upper bound. The second sum is clearly bounded above by 1. Hence, (4.3) holds.                                            □

## 4.3 Applications of Theorem 4.1.1

### 4.3.1 Covering the $n$-dimensional sphere

As a first application of Theorem 4.1.1 we consider the problem of covering the $n$-dimensional sphere

$$X = S^n = \{x \in \mathbb{R}^{n+1} : x \cdot x = 1\},$$

equipped with spherical distance

$$d(x, y) = \arccos x \cdot y \in [0, \pi]$$

and with the rotationally invariant probability measure $\omega$, by spherical caps / metric balls $B(x, r)$. Clearly, properties (a) and (b) are satisfied in this setting. Again we set $\omega_r = \omega(B(x, r))$.
We are especially interested in the covering number $N(S^n, r)$ when $0 < r < \pi/2$ or equivalently in the covering density defined by $\omega_r \cdot N(S^n, r)$. Theorem 4.1.1 says that the covering density is at most

$$\frac{\omega_r}{\omega_{r-\varepsilon}}\left(\ln\left(\frac{\omega_{r-\varepsilon}}{\omega_\varepsilon}\right) + 1\right). \tag{4.6}$$

This upper bounds holds for every $\varepsilon$ with $0 < \varepsilon < r$. By choosing $\varepsilon$ depending on the dimension $n$ and on the spherical distance $r$ we can find an upper bound for the covering density which only depends on $n$.

For this we recall a useful estimate of fractions of the form $\omega_{tr}/\omega_r$ due to Börözky Jr. and Wintsche [11]:

$$\frac{\omega_{tr}}{\omega_r} \leq t^n \quad \text{whenever } r < tr < \frac{\pi}{2}. \tag{4.7}$$

We set $\varepsilon = r/(\mu n + 1)$ with parameter $\mu > 1$ which we are going to adjust later. Furthermore, we set

$$t = \frac{r}{r - \varepsilon} = 1 + \frac{1}{\mu n}$$

and

$$t' = \frac{r - \varepsilon}{\varepsilon} = \mu n.$$

By using (4.6) and (4.7) we have the following upper bound for the covering density

$$\frac{\omega_r}{\omega_{r-\varepsilon}} \left( \ln \left( \frac{\omega_{r-\varepsilon}}{\omega_\varepsilon} \right) + 1 \right) \leq \left( 1 + \frac{1}{\mu n} \right)^n (n \ln \mu n + 1)$$

$$\leq e^{1/\mu}(n \ln \mu n + 1)$$

$$\leq \left( 1 + \frac{1}{\mu - 1} \right)(n \ln \mu n + 1)$$

Thus we have proven:

**Corollary 4.3.1.** *The covering density of the n-dimensional sphere by spherical balls is at most*

$$\left( 1 + \frac{1}{\mu - 1} \right)(n \ln \mu n + 1) \text{ for all } \mu > 1.$$

*In particular, for $\mu = \ln n$, the covering density is at most*

$$n \ln n + n \ln \ln n + n + o(n).$$

In the asymptotic case the best known bound is $(1/2 + o(1))n \ln n$ due to Dumer [29] which comes from a randomized construction. Our corollary slightly improves the previously best known non-asymptotic bound $n \ln n + n \ln \ln n + 2n + o(n)$ by Börözky Jr. and Wintsche [11] also coming from a randomized construction.

## 4.3.2   Covering $n$-dimensional Euclidean space

As a second application we consider coverings of $n$-dimensional Euclidean space $\mathbb{R}^n$ by congruent balls. We get a covering of $\mathbb{R}^n$ by applying Theorem 4.1.1 to the torus $\mathbb{T}^n =$

$\mathbb{R}^n/\mathbb{Z}^n$ which is a compact metric space satisfying properties (a) and (b). Then we periodically extend the obtained covering of $\mathbb{T}^n$ to a covering of the entire $\mathbb{R}^n$ having the same covering density.

We repeat the choices and calculations as in the previous section (which are slightly simpler here because clearly $\omega_{tr}/\omega_r = t^n$ holds where here $\omega$ denotes the Lebesgues measure) and get:

**Corollary 4.3.2.** *The covering density of the n-dimensional Euclidean space by congruent balls is at most*

$$\left(1 + \frac{1}{\mu - 1}\right)(n \ln \mu n + 1) \text{ for all } \mu > 1.$$

*In particular, for $\mu = \ln n$, the covering density is at most*

$$n \ln n + n \ln \ln n + n + o(n).$$

We remark that this bound coincides with the currently best known bound by Féjes Toth [30] coming from a deterministic construction. The best known bound coming from a randomized construction is $(1/2 + o(1))n \ln n$ due to Dumer [29]

### 4.3.3   More general coverings

At last we want to demonstrate that the greedy approach to geometric covering problems is quite flexible. It is not restricted to finding coverings of compact metric spaces by balls. With small modifications it can for example be applied to prove the following theorem due to Naszódi [60, Theorem 1.3]:

**Theorem 4.3.3.** *Let $K \subseteq \mathbb{R}^n$ be a bounded measurable set. Then there is a covering of $\mathbb{R}^n$ by translated copies of K of density at most*

$$\inf\left\{\frac{\omega(K)}{\omega(K_{-\delta})}\left(\ln\left(\frac{\omega(K_{-\delta/2})}{\omega(B(0, \delta/2))}\right) + 1\right) : \delta > 0\right\},$$

*where $K_{-\delta} = \{x \in K : B(x, \delta) \subseteq K\}$ is the δ-inner parallel body of K.*

Here, we only sketch the proof, though filling in the details can be done with the theory illustrated in Chapter 2. As in Section 4.3.2 we can work on the torus $\mathbb{T}^n$. We approximate the body $K$ and its inner parallel bodies by a finite union of balls. For this we choose points $y_1, \ldots, y_N$ such that

$$K_{-\delta} \subseteq \bigcup_{i=1}^{N} B(y_i, \delta/2) \subseteq K_{-\delta/2}$$

and

$$\bigcup_{i=1}^{N} B(y_i, \delta) \subseteq K$$

holds. We want to cover $\mathbb{T}^n$ greedily using these unions. Whereas in the greedy algorithm we use the centers of the metric balls to indicate which balls we picked, we use the translation vectors translating the union of balls here.

We make this statement precise in the general setting of a compact metric space $(X, d)$. Consider the group of continuous isometries of $(X, d)$, these are all continuous bijective maps $\tau: X \rightarrow X$ which preserve the distance between every two points $x, y \in X$. We assume that the group acts transitively on $X$ and that $\omega(\tau A) = \omega(A)$ holds for all continuous isometries $\tau$ and all measurable sets $A$. Then by the theorem of Arzelà-Ascoli (see for example [33, Chapter 4.6]) the group of continuous isometries is relatively compact in the compact space of continuous maps mapping $X$ to itself equipped with the supremum norm. We need this compactness for Lemma 4.2.1. So we can transfer the analysis of the greedy algorithm given in Section 4.2 to this setting.

In the end going back from the torus $\mathbb{T}^n$ to $\mathbb{R}^n$ we get a covering of $\mathbb{R}^n$ with translated copies of $K$ with density at most

$$\inf\left\{\frac{\omega(K)}{\omega(K_{-\delta/2})}\left(\ln\left(\frac{\omega\left(K_{-\delta/2}\right)}{\omega(B(0,\delta/2))}\right) + 1\right) : \delta > 0\right\},$$

improving the result of Naszódi slightly.

# Acknowledgements

We thank Markus Schweighofer and Cordian Riener for helpful remarks.

# CHAPTER FIVE

# Lower bounds

This chapter is based on the publication *An SDP hierarchy for the covering problem*, by D. de Laat, C. Riener, J. H. Rolfes and F. Vallentin, in preparation.

We provide lower bounds for the covering number $N(X, r)$ on compact metric spaces $(X, d)$. This problem can be seen as an infinite instance of SET COVER, where the subsets $S_i \subseteq X$ are determined by the metric $d$ and the scalar $r \in \mathbb{R}_{\geq 0}$. We give a sequence of lower bounds for this problem inspired by Lasserre's hierarchy of semidefinite programs for SET COVER. Additionally, we develop a duality theory for these programs and prove that they converge to $N(X, r)$ after finitely many steps. For the particular case that $X$ is the two-dimensional unit sphere $S^2$, we derive a finite-dimensional semidefinite program that provides lower bounds for $N(S^2, r)$.

## 5.1 Introduction

We recall the definition of the covering problem as stated in Chapter 1: Let $X$ be a compact metric space having metric $d$. Given a scalar $r \in \mathbb{R}_{\geq 0}$ we define the *closed ball* of radius $r$ around center $x \in X$ by

$$B(x, r) = \{y \in X : d(x, y) \leq r\}.$$

The *covering number* of the space $X$ and a positive number $r$ is the smallest number of such balls with radius $r$ one needs to cover $X$, i.e.,

$$N(X, r) = \min\left\{|Y| : Y \subseteq X, \bigcup_{y \in Y} B(y, r) = X\right\}. \tag{5.1}$$

Determining the covering number is a fundamental problem in metric geometry (see for example the classical book by Rogers [68]). Among others, applications arise in the field

of compressive sensing [34], approximation theory and machine learning [17] or, as illustrated in Chapter 1, in probability theory [55] and theoretical quantum computing [61].
So far, there exist upper bounds for the covering number of several specific metric spaces. For a survey on this, we refer to Naszódi [59]. We recall the following recent result presented in Chapter 4. It provides upper bounds for the covering number for a wide class of compact metric spaces. Suppose $X$ is equipped with a probability measure $\omega$ satisfying the following two conditions:

(a) $\omega(B(x, s)) = \omega(B(y, s))$ for all $x, y \in X$, and for all $s \geq 0$,

(b) $\omega(B(x, \varepsilon)) > 0$ for all $x \in X$, and for all $\varepsilon > 0$.

By (a) the measure of a ball does only depend on the radius $s$ and not on the center $x$, so we simply denote $\omega(B(x, s))$ by $\omega_s$ and obtain for every $\varepsilon$ with $r/2 > \varepsilon > 0$ the following bounds on the covering number:

$$\frac{1}{\omega_r} \leq \mathcal{N}(X, r) \leq \frac{1}{\omega_{r-\varepsilon}}\left(\ln\left(\frac{\omega_{r-\varepsilon}}{\omega_\varepsilon}\right) + 1\right). \tag{5.2}$$

The trivial lower bound above is known as the *volume bound*. For coverings of the (non-compact) Euclidean space we further have the celebrated *Coxeter-Few-Rogers bound* [16] of which Böröczky and Wintsche gave an adjusted version for the case of $X = S^n$, i.e., the unit sphere equipped with the spherical distance $d(x, y) := \arccos(x \cdot y)$. Their bound states that for sufficiently small radii $r$ we can bound the covering number by

$$\frac{c \cdot n}{\omega_r} \leq \mathcal{N}(S^n, r)$$

from below, where $c$ is an absolute constant. This is the best lower bound known so far for the spherical case except for Fejes-Tóth's [31] bound for the case $n = 2$, which is relying on similar techniques.

The main contribution of the present chapter is the construction of a unified framework in order to compute lower bounds for a wide class of covering numbers. More concretely, we define a hierarchy $\mathcal{N}^t(X, r)$ of increasingly strong relaxations of the covering problem on $X$, where each instance is a conic minimization problem. The approach used to design these relaxations is strongly inspired by the moment method of Lasserre [54] for finite 0/1-integer programs and his seminal paper [53] on general polynomial optimization problems. We suppose that $X$ is *homogeneous* with respect to its isometry group $\Gamma$, i.e., $\Gamma$ acts transitively on $X$ and the covering constraints are indexed by $\gamma \in \Gamma$ instead of $x \in X$. We then show that $\mathcal{N}^t(X, r)$ yields an increasing, converging sequence of lower bounds

$$\frac{1}{\omega_r} = \mathcal{N}^1(X, r) \leq \mathcal{N}^2(X, r) \leq \ldots \leq \mathcal{N}^\alpha(X, r) = \mathcal{N}(X, r),$$

where $\alpha \in \mathbb{N}$ is a natural number determined by the specific instance of the problem. Additionally, we exploit the symmetry that arises from $\Gamma$ acting on $X$ to achieve symmetrized reformulations of $\mathcal{N}^t(X, r)$ and use this symmetry to determine a finite-dimensional semidefinite program, which is a good candidate to approximate $\mathcal{N}^2(S^2, r)$ upto arbitrary precision.

## 5.2   A semidefinite programming hierarchy

### 5.2.1   The covering number as an optimization problem

In the following paragraphs we aim to restate the covering number as an optimization problem using Dirac measures. The intuition to do this arises from graph theory, in particular from a variant of the SET COVER problem called HITTING SET: For a set $V$ we define for each element $u \in V$ corresponding sets $S_u \subseteq V$ and consider the following integer optimization problem

$$SC := \min\left\{\sum_{v \in V} x_v : \sum_{v \in S_u} x_v \geq 1 \text{ for every } u \in V, \ x \in \{0, 1\}^V\right\}. \tag{5.3}$$

As was described in Chapter 3, Lasserre [54] introduced an approximation technique for such 0/1-integer programs, where he lifted the vectors $x \in \mathbb{R}^V$ to vectors $y \in \mathbb{R}^{V_{2t}}$, with $V_{2t} := \{U \subseteq V : |U| \leq 2t\}$. On the lifted space one can now solve an approximate problem for every $t$, which provides increasingly better lower bounds for $SC$ with growing $t$. This technique has been successfully applied to *packing problems* [23] and *energy minimization problems* [21].

The main goal of this chapter is to define a similar sequence $\mathcal{N}^t(X, r)$ of convex optimization problems, that also provides good lower bounds for $\mathcal{N}(X, r)$ by lifting coverings to a higher dimensional space. In the next paragraphs we define this space for uncountable sets $V$.

Although one could interpret $y \in \mathbb{R}^{V_{2t}}$ as functions from $V_{2t}$ to $\mathbb{R}$ whenever $V$ is uncountable, it is not clear how to define a suitable objective function in this case. Instead we consider measures $\mu$ defined on the sigma-algebra $\mathcal{B}(X)$ of Borel sets of $X$ to define a similar program to (5.3) and deal with the objective function. We handle the uncountable number of constraints by assuming a transitive group structure, i.e., we depict $\mathcal{N}(X, r)$ as

$$\mathcal{N}(X, r) = \inf\left\{\mu(X) : \mu = \sum_{y \in Y} \delta_y, \ \mu(B(\gamma x, r)) \geq 1 \text{ for all } \gamma \in \Gamma\right\},$$

where $\delta_y : \mathcal{B}(X) \to \mathbb{R}$ is the Dirac measure defined by

$$\delta_y(A) = \begin{cases} 1 & \text{if } y \in A \\ 0 & \text{otherwise.} \end{cases}$$

We further recall that the finite sums of Dirac measures lie in the cone $\mathcal{M}(\mathcal{B}(X))_{\geq 0}$ of nonnegative Radon measures on $\mathcal{B}(X)$ – similar to the vectors $y \in \mathbb{R}^{V_{2n}}$ defined on the power set $V_{2n} = \mathcal{P}([n])$ of $[n]$ in the $n$-th step of Lasserre's hierarchy. In the next section we define the infinite-dimensional counterparts of the sets $V_{2t}$.

## 5.2.2   Approximating the cone of Radon measures

For an inner approximation of the cone $\mathcal{M}(\mathcal{B}(X))_{\geq 0}$ we restrict the ground set $\mathcal{B}(X)$ to the set of subsets $X_t := \{Y \subseteq X : |Y| \leq t\}$ of cardinality at most $t$. In fact, after proving some useful topological properties for $X_t$, we will restrict $\mathcal{B}(X)$ even further to a set $I_t \subseteq X_t \subseteq \mathcal{B}(X)$. But first, we equip the non-empty sets in $X_t$ with the *Hausdorff distance*

$$d_H(Y, Z) := \max \left\{ \sup_{y \in Y} \inf_{z \in Z} d(y, z), \sup_{z \in Z} \inf_{y \in Y} d(y, z) \right\}$$

and extend this distance to a metric

$$d'_H(Y, Z) := \begin{cases} d_H(Y, Z) & \text{if } Y, Z \neq \emptyset \\ 0 & \text{if } Y = Z = \emptyset \\ \sup_{x, y \in X} d(x, y) & \text{otherwise,} \end{cases}$$

which also takes care of $\emptyset \in X_t$. This implies that $X_t$ is Hausdorff. To show compactness of $X_t$ we consider the continuous map

$$q : \bigoplus_{i=1}^{t} X \to (X_t, d_H), \text{ defined by } q(x_1, \ldots, x_t) = \{x_1, \ldots x_t\}$$

equipped with the product metric. Thus we have that $X_t = Im(q) \cup \{\emptyset\}$ is compact. Due to the Hausdorff property of $X$ we know that for every optimal solution $Y^* \subseteq X$ of (5.1) and $y, y' \in Y^*$ we have that $d(y, y') \geq \varepsilon > 0$. If we choose

$$\varepsilon = \min\{d(y, y') : y, y' \in Y^*, y \neq y'\},$$

we can restrict $X_t$ even further to the space

$$I_t := \{Y \subseteq X : |Y| \leq t, \ d(x, y) \geq \varepsilon \text{ for all } x, y \in Y\},$$

and still have ensured that at least one optimal $Y^* \in X_t$ is contained in $I_t$. The concrete value of this $\varepsilon$ is a priori not known. However, for every $\varepsilon > 0$, as a subset of $X_t$, $I_t$ inherits the Hausdorff property, whereas compactness is shown in the following Lemma 5.2.1, a specified version of a lemma in [23]. Since [23] considers topological spaces instead of metric spaces, our proof is slightly simpler by being able to exploit the additional structure given by the metric. In both proofs, the set

$$(U_1, \ldots, U_k)_t := \{Y \in X_t : Y \subseteq U_1 \cup \ldots \cup U_k, \ Y \cap U_i \neq \emptyset \text{ for } 1 \leq i \leq k\} \tag{5.4}$$

plays a significant role.

**Lemma 5.2.1.** *[23] The set $I_t$ is compact for every $t \in \mathbb{N}$ with respect to the topology induced by the metric $d'_H$.*

*Proof.* As a subset of the compact space $X_t$, $I_t$ is totally bounded and complete if and only if it is closed. We show that $I_t$ is closed by considering $\{x_1, \ldots x_k\} \in X_t \setminus I_t$ for pairwise disjoint $x_i$. Then without loss of generality $d(x_1, x_2) < \varepsilon$ holds, implying that there is an open neighbourhood $B_\varepsilon$ such that $x_1, x_2 \in B_\varepsilon$, e.g., $B_\varepsilon = B(x_1, \varepsilon)$. The Hausdorff property implies that there are disjoint open neighbourhoods $U_1, U_2$ for which $x_1 \in U_1 \subseteq B_\varepsilon$, $x_2 \in U_2 \subseteq B_\varepsilon$ holds. Thus $(U_1, U_2, X, \ldots, X)_t \notin I_t$ is an open neighbourhood of $\{x_1, \ldots, x_k\}$ implying that $X_t \setminus I_t$ is open.     □

In the remainder of this chapter we will focus on the compact metric space $(I_t, d'_H)$ and its corresponding cone $\mathcal{M}(I_t)_{\geq 0}$ to approximate the covering number. We further denote $B_r \subseteq X$ as a fixed $r$-ball in $X$ and assume that the group $\Gamma$ is a transitive isometry group of $X$. Examples of such covering problems are the space $(S^n, d)$ with spherical distance $d$ and isometry group $\Gamma = O(n)$ or the flat torus equipped with the group of translations.

### 5.2.3   Continuous Lasserre-hierarchy

We first introduce some notations and basic facts in order to define our covering hierarchy. The set $C(I_t)$ of *continuous real-valued functions* on $I_t$ and the set $\mathcal{M}(I_t)$ of *signed Radon measures* form a topological dual pairing, where the topology is defined by the supremum norm on $C(I_t)$. This is a consequence of the Riesz representation theorem (see, e.g., [82]). In particular, these dual spaces contain the cones $C(I_t)_{\geq 0}$ of nonnegative functions in $C(I_t)$ and its dual cone

$$\mathcal{M}(I_t)_{\geq 0} := \left\{ \mu \in \mathcal{M}(I_t) : \int_{I_t} f d\mu \geq 0 \text{ for all } f \in C(I_t)_{\geq 0} \right\}.$$

Similarly, we denote the space of *symmetric kernels* $C(I_t \times I_t)_{sym}$ as the set of continuous functions $K : I_t \times I_t \to \mathbb{R}$, where $K(J, J') = K(J', J)$ for any $J, J' \in I_t$. Such a kernel is referred to as *positive semidefinite* if the matrices $(K(J_k, J_l))_{k,l}$ are positive semidefinite for any finite choice of sets $J_1, \ldots J_m \in I_t$. The cone of all these semidefinite kernels is denoted by $C(I_t \times I_t)_{\geq 0}$, its dual cone is denoted by $\mathcal{M}(I_t \times I_t)_{\geq 0}$ and is contained in the space of *symmetric Radon measures* $\mathcal{M}(I_t \times I_t)_{sym.}$, where $\mu(E, E') = \mu(E', E)$ holds for every pair of Borel sets $E, E'$. The continuous hierarchy which relaxes $\mathcal{N}(X, r)$ is defined as follows:

**Definition 5.2.2.** *For $t \in \mathbb{N}$ and $I_{=t} = \{J \in I_t : |J| = t\}$ we define*

$$\mathcal{N}^t(X, r) := \inf \mu(I_{=1}) \tag{5.5}$$

$$\mu \in \mathcal{M}(I_{2t})_{\geq 0},$$
$$\mu(\{\emptyset\}) = 1,$$
$$A_t \mu \in \mathcal{M}(I_t \times I_t)_{\geq 0},$$
$$A_t^\gamma \mu \in \mathcal{M}(I_{t-1} \times I_{t-1})_{\geq 0} \text{ for all } \gamma \in \Gamma,$$

*where the operators $A_t : \mathcal{M}(I_{2t}) \to \mathcal{M}(I_t \times I_t)_{sym}$ and $A_t^\gamma : \mathcal{M}(I_{2t}) \to \mathcal{M}(I_{t-1} \times I_{t-1})_{sym}$ are defined pointwise by their* adjoint *operators $B_t : C(I_t \times I_t)_{sym} \to C(I_{2t})$ and $B_t^\gamma : C(I_{t-1} \times I_{t-1})_{sym} \to C(I_{2t})$, i.e., by the unique operators that satisfy $\langle A_t \mu, K \rangle = \langle \mu, B_t K \rangle$ respectively $\langle A_t^\gamma \mu, K \rangle = \langle \mu, B_t^\gamma K \rangle$. We define those operators by*

$$B_t K(S) := \sum_{J, J' \in I_t:\ J \cup J' = S} K(J, J') \qquad and \tag{5.6}$$

$$B_t^\gamma K(S) := \sum_{x \in \gamma B_r\ J, J' \in I_{t-1}:\ J \cup J' \cup \{x\} = S} K(J, J') - \sum_{J, J' \in I_{t-1}:\ J \cup J' = S} K(J, J'). \tag{5.7}$$

**Remark 5.2.3.** *We observe that $\|B_t K\|_\infty \leq 2^{2t} \|K\|_\infty$ and*

$$\|B_t^\gamma K\|_\infty \leq 2t 2^{2(t-1)} \|K\|_\infty + 2^{2(t-1)} \|K\|_\infty$$

*are bounded and hence continuous. Thus we have proven the existence of $A_t$ and $A_t^\gamma$.*

Increasing $t$ strengthens the bound by imposing more constraints to the feasible region, to be precise: if $\mu_{t+1} \in \mathcal{M}(I_{2t+2})$ is feasible for $\mathcal{N}^{t+1}(X, r)$ then $\mu_t := \mu_{t+1}\big|_{I_{2t}}$ is feasible for $\mathcal{N}^t(X, r)$ with the same objective value. Additionally, the measure $\mu^* := \sum_{Z \in I_{2t}:\ Z \subseteq Y} \delta_Z$ is feasible for every step in our hierarchy whenever the set $Y$ is a finite covering of $X$. Thus we have:

$$\mu^*(\{\emptyset\}) = \sum_{Z \in I_{2t}:\ Z \subseteq Y} \delta_Z(\{\emptyset\}) = \delta_\emptyset(\{\emptyset\}) = 1,$$

and additionally, for every $K \in C(I_t \times I_t)_{\geq 0}$:

$$\langle A_t \mu^*, K \rangle = \langle \mu^*, B_t K \rangle$$

$$= \sum_{Z \in I_{2t}: Z \subseteq Y} \sum_{J,J' \in I_t, \ Z=J \cup J'} K(J, J')$$

$$= \sum_{J,J' \in I_t: \ J,J' \subseteq Y} K(J, J') \geq 0,$$

$$\langle A_t^\gamma \mu^*, K \rangle = \langle \mu^*, B_t^\gamma K \rangle$$

$$= \sum_{Z \in I_{2t}: \ Z \subseteq Y \ x \in \gamma B_r, \ J,J' \in I_{t-1}, \ Z=J \cup J' \cup \{x\}} \sum K(J, J')$$

$$- \sum_{J,J' \in I_{t-1}: \ J,J' \subseteq Y} K(J, J')$$

$$= \sum_{x \in Y \cap \gamma B_r, \ J,J' \in I_{t-1}: \ J,J' \subseteq Y} \sum K(J, J') - \sum_{J,J' \in I_{t-1}: \ J,J' \subseteq Y} K(J, J') \geq 0,$$

where the last inequality is implied by the fact that $Y \cap \gamma B_r \neq \emptyset$ since $Y$ is a covering. The objective value is $\mu^*(I_{=1}) = |Y|$ and thus yields the covering number if $Y$ is a minimal covering. Hence we have shown the following theorem:

**Theorem 5.2.4.** *With the notation defined in Definition 5.2.2 we have the following non-decreasing sequence of bounds*

$$\mathcal{N}^1(X, r) \leq \mathcal{N}^2(X, r) \leq \ldots \leq \mathcal{N}^t(X, r) \leq \mathcal{N}(X, r)$$

*for any value of t.*

We compare this sequence to the bounds from Theorem 4.1.1. Suppose we have a probability measure $\omega$ satisfying properties (a) and (b), then in the case $t = 1$ we observe that the measure

$$\mu'(A) := \begin{cases} 1 & \text{if } A = \{\emptyset\}, \\ \frac{\omega(\bigcup_{a \in A} a)}{\omega_r} & \text{if } A \subseteq I_{=1}, \\ 0 & \text{otherwise.} \end{cases}$$

is a feasible measure for $\mathcal{N}^1(X, r)$. Its objective value is $\frac{1}{\omega_r}$ and thus implies

$$\mathcal{N}^1(X, r) \leq \frac{1}{\omega_r}. \tag{5.8}$$

Furthermore, we shall prove Theorem 5.4.3, which states that in fact even equality holds. Thus we retrieve the volume bound

$$\frac{1}{\omega_r} = \mathcal{N}^1(X, r) \leq \mathcal{N}(X, r)$$

in the first step of our hierarchy.

## 5.3  Symmetry Reduction

In this section we will reduce the hierarchy (5.5) defined on the space of signed Radon measures $\mathcal{M}(I_{2t})$ first, to a hierarchy acting on the space of $\Gamma$-invariant signed Radon measures and second, to Radon measures $\mu \in \mathcal{M}(I_{2t}/\Gamma)$ acting on the quotient space $I_{2t}/\Gamma$. Ultimately, this will reduce the potentially infinite number of covering constraints $A_t^\gamma \mu \in \mathcal{M}(I_{t-1} \times I_{t-1})_{\geq 0}$ to a single constraint $A_t^{id} \mu \in \mathcal{M}(I_{t-1} \times I_{t-1})_{\geq 0}$.

We call a measure $\mu : \mathcal{B}(I_t) \to \mathbb{R}$, $\Gamma$-*invariant* if and only if $\mu(\gamma S) = \mu(S)$ for every $S \in \mathcal{B}(I_t)$ and $\gamma \in \Gamma$ acting pointwise on the elements in $S$, i.e., $\gamma\{s_1, \ldots, s_t\} = \{\gamma s_1, \ldots, \gamma s_t\}$ for $S = \{s_1, \ldots, s_t\}$. We denote these $\Gamma$-invariant, signed Radon measures by $\mathcal{M}(I_t)^\Gamma$. In this space we can define the cone

$$\mathcal{M}(I_t)_{\geq 0}^\Gamma := \mathcal{M}(I_t)_{\geq 0} \cap \mathcal{M}(I_t)^\Gamma.$$

To apply symmetry reduction for a reformulation of $N^t(X, r)$, we need the following lemma.

**Lemma 5.3.1.** *Suppose* $\mu \in \mathcal{M}(I_{2t})_{\geq 0}^\Gamma$. *Then we have*

$$A_t^{id} \mu \in \mathcal{M}(I_{t-1} \times I_{t-1})_{\geq 0} \iff A_t^\gamma \mu \in \mathcal{M}(I_{t-1} \times I_{t-1})_{\geq 0} \text{ for every } \gamma \in \Gamma.$$

*Proof.* The "$\Leftarrow$" implication is immediate by considering $\gamma = id$.

For the "$\Rightarrow$" implication we consider for an arbitrary kernel $K \in C(I_{t-1} \times I_{t-1})_{\geq 0}$ the shifted kernel $K^\gamma \in C(I_{t-1} \times I_{t-1})_{\geq 0}$ defined by $K^\gamma(J, J') = K(\gamma J, \gamma J')$ and observe the following:

$$\langle A_t^\gamma \mu, K \rangle = \langle \mu, B_t^\gamma K \rangle = \int_{S \in I_{2t}} B_t^\gamma K(S)\,d\mu(S)$$

$$= \int_{S \in I_{2t}} \sum_{\substack{x \in \gamma B_r,\ J, J' \in I_{t-1}:\ J \cup J' \cup \{x\} = S}} K(J, J') - \sum_{\substack{J, J' \in I_{t-1}:\ J \cup J' = S}} K(J, J')\,d\mu(S)$$

$$= \int_{S \in I_{2t}} \sum_{\substack{\gamma x \in \gamma B_r,\ \gamma J, \gamma J' \in I_{t-1}:\\ \gamma J \cup \gamma J' \cup \gamma\{x\} = S}} K(\gamma J, \gamma J') - \sum_{\substack{\gamma J, \gamma J' \in I_{t-1}:\ \gamma J \cup \gamma J' = S}} K(\gamma J, \gamma J')\,d\mu(S)$$

$$= \int_{S \in I_{2t}} \sum_{\substack{x \in B_r,\ J, J' \in I_{t-1}:\ J \cup J' \cup \{x\} = \gamma^{-1} S}} K^\gamma(J, J') - \sum_{\substack{J, J' \in I_{t-1}:\ J \cup J' = \gamma^{-1} S}} K^\gamma(J, J')\,d\mu(S)$$

$$= \int_{S \in I_{2t}} B_t^{id} K^\gamma(\gamma^{-1} S)\,d\mu(S) \overset{\mu \in \mathcal{M}(I_{2t})_{\geq 0}^\Gamma}{=} \langle \mu, B_t^{id} K^\gamma \rangle = \langle A_t^{id} \mu, K^\gamma \rangle.$$

Thus we have that $\langle A_t^\gamma \mu, K \rangle \geq 0$ for every $K \in C(I_{t-1} \times I_{t-1})_{\geq 0}$ whenever $\langle A_t^{id} \mu, K \rangle \geq 0$ for every $K \in C(I_{t-1} \times I_{t-1})_{\geq 0}$, implying the claim. $\qquad \square$

Since $\Gamma$ is a compact topological group, there exists a normalized Haar measure $\lambda$ on this group (see, e.g., [24] for a proof). The corresponding Haar integral turns out to be a convenient counterpart to the averaging operator in the finite case, i.e., it helps us to reformulate the program (5.5) as follows:

**Theorem 5.3.2.** *Let $t \in \mathbb{N}$ and $I_{=t} = \{J \in I_t : |J| = t\}$, then*

$$\mathcal{N}^t(X, r) = \inf \mu(I_{=1}) \tag{5.9}$$

$$\mu \in \mathcal{M}(I_{2t})_{\geq 0}^{\Gamma},$$
$$\mu(\{\emptyset\}) = 1,$$
$$A_t \mu \in \mathcal{M}(I_t \times I_t)_{\geq 0},$$
$$A_t^{id} \mu \in \mathcal{M}(I_{t-1} \times I_{t-1})_{\geq 0}.$$

*Proof.* First, we show that for every feasible measure $\mu$, the measure $\bar{\mu} \in \mathcal{M}(I_{2t})_{\geq 0}^{\Gamma}$ defined by

$$\bar{\mu}(A) := \int_{\Gamma} \mu(\gamma A) d\lambda(\gamma)$$

is feasible for $\mathcal{N}^t(X, r)$ with the same objective value. Trivially, we have that

$$\bar{\mu}(I_{=1}) = \mu(I_{=1}), \quad \bar{\mu}(\{\emptyset\}) = \mu(\{\emptyset\}) = 1.$$

For the remaining constraints, two identities that follow immediately from the definition of $B_t$ and $B_t^{\gamma}$, are very useful: For a symmetric kernel $K$ we consider again the shifted kernel $K^{\gamma}$, i.e., $K^{\gamma}(J, J') = K(\gamma J, \gamma J')$, and observe that

$$B_t K(\gamma S) = B_t K^{\gamma}(S) \text{ and } B_t^{\gamma} K(\gamma S) = B_t^{\gamma^{-1}\gamma} K^{\gamma}(S),$$

where the latter identity follows from

$$B_t^{\gamma} K(\gamma S) = \sum_{x \in \gamma' B_r,\ J,J' \in I_{t-1}:\ J \cup J' \cup \{x\} = \gamma S} K(J, J') - \sum_{J,J' \in I_{t-1}:\ J \cup J' = \gamma S} K(J, J')$$

$$= \sum_{\gamma x \in \gamma' B_r,\ \gamma J, \gamma J' \in I_{t-1}:\ \gamma J \cup \gamma J' \cup \gamma \{x\} = \gamma S} K(\gamma J, \gamma J') - \sum_{\gamma J, \gamma J' \in I_{t-1}:\ \gamma J \cup \gamma J' = \gamma S} K(\gamma J, \gamma J')$$

$$= \sum_{x \in \gamma^{-1}\gamma' B_r,\ J,J' \in I_{t-1}:\ J \cup J' \cup \{x\} = S} K^{\gamma}(J, J') - \sum_{J,J' \in I_{t-1}:\ J \cup J' = S} K^{\gamma}(J, J')$$

$$= B_t^{\gamma^{-1}\gamma} K^{\gamma}(S).$$

Together with the property $K^{\gamma} \geq 0 \iff K \geq 0$ this leads to

$$\langle A_t \bar{\mu}, K \rangle = \langle \bar{\mu}, B_t K \rangle = \int_{\Gamma} \int_{S \in I_{2t}} B_t K(S) d\mu(\gamma S) d\lambda(\gamma)$$

$$= \int_{\Gamma} \int_{S \in I_{2t}} B_t K(\gamma^{-1} S) d\mu(S) d\lambda(\gamma)$$

$$\overset{B_t K(\gamma S) = B_t K^{\gamma}(S)}{=} \int_{\Gamma} \int_{S \in I_{2t}} B_t K^{\gamma^{-1}}(S) d\mu(S) d\lambda(\gamma)$$

$$= \int_{\Gamma} \langle \mu, B_t K^{\gamma^{-1}} \rangle d\lambda(\gamma) = \int_{\Gamma} \langle A_t \mu, K^{\gamma^{-1}} \rangle d\lambda(\gamma) \geq 0,$$

and

$$\langle A_t^\gamma \bar\mu, K \rangle = \langle \bar\mu, B_t^\gamma K \rangle = \int_\Gamma \int_{S \in I_{2t}} B_t^\gamma K(S) d\mu(\gamma S) d\lambda(\gamma)$$

$$= \int_\Gamma \int_{S \in I_{2t}} B_t^\gamma K(\gamma^{-1} S) d\mu(S) d\lambda(\gamma)$$

$$\overset{B_t^\gamma K(\gamma S) = B_t^{\gamma^{-1}\gamma'} K^{\gamma'}(S)}{=} \int_\Gamma \int_{S \in I_{2t}} B_t^{\gamma\gamma'} K^{\gamma^{-1}}(S) d\mu(S) d\lambda(\gamma)$$

$$= \int_\Gamma \langle \mu, B_t^{\gamma\gamma'} K^{\gamma^{-1}} \rangle d\lambda(\gamma) = \int_\Gamma \langle A_t^{\gamma\gamma'} \mu, K^{\gamma^{-1}} \rangle d\lambda(\gamma) \geq 0.$$

Thus we can restrict our program $\mathcal{N}^t(X, r)$ to the cone $\mathcal{M}(I_{2t})_{\geq 0}^\Gamma$ and we can apply Lemma 5.3.1 to obtain

$$\mathcal{N}^t(X, r) = \inf \mu(I_{=1}) \tag{5.10}$$

$$\mu \in \mathcal{M}(I_{2t})_{\geq 0}^\Gamma,$$
$$\mu(\{\emptyset\}) = 1,$$
$$A_t\mu \in \mathcal{M}(I_t \times I_t)_{\geq 0},$$
$$A_t^{id}\mu \in \mathcal{M}(I_{t-1} \times I_{t-1})_{\geq 0}.$$

$\square$

However, we would like to work with dual cones $C(V)_{\geq 0}$ and $\mathcal{M}(V)_{\geq 0}$ on a compact Hausdorff space $V$ to incorporate these cones into a duality theory. In the upcoming paragraphs of this chapter we will address the constraint that the measure has to be $\Gamma$-invariant by restricting the cone $\mathcal{M}(I_{2t})_{\geq 0}$ to the cone $\mathcal{M}(I_{2t}/\Gamma)_{\geq 0}$ over the quotient space $I_{2t}/\Gamma$, which is a compact Hausdorff space as well. Eventually, we will derive the following formulation:

**Theorem 5.3.3.**

$$\mathcal{N}^t(X, r) = \inf \mu(I_{=1}) \tag{5.11}$$

$$\mu \in \mathcal{M}(I_{2t}/\Gamma)_{\geq 0},$$
$$\mu(\{\emptyset\}) = 1,$$
$$\tilde{A}_t\mu \in \mathcal{M}(I_t \times I_t)_{\geq 0},$$
$$\tilde{A}_t^{id}\mu \in \mathcal{M}(I_{t-1} \times I_{t-1})_{\geq 0},$$

where $\tilde{A}_t$ and $\tilde{A}_t^{id}$ are defined pointwise by their adjoint operators $\tilde{B}_t : C(I_t \times I_t) \to C(I_{2t}/\Gamma)$ and $\tilde{B}_t^{id} : C(I_{t-1} \times I_{t-1}) \to C(I_{2t}/\Gamma)$ defined by

$$\tilde{B}_t K(\pi(S)) := \int_\Gamma \sum_{J, J' \in I_t: \ J \cup J' = \gamma S} K(J, J') d\lambda(\gamma)$$

$$\tilde{B}_t^{id} K(\pi(S)) := \int_\Gamma \sum_{\substack{x \in B_r \; J,J' \in I_{t-1}: \; J \cup J' \cup \{x\} = \gamma S}} K(J, J') - \sum_{\substack{J,J' \in I_{t-1}: \; J \cup J' = \gamma S}} K(J, J') d\lambda(\gamma),$$

where $\lambda$ is the normalized Haar measure on $\Gamma$ and $\pi : I_t \to I_t/\Gamma$ the quotient map.

We first observe that

$$\|\tilde{B}_t K\|_\infty \le \int_\Gamma \|B_t K\|_\infty d\lambda(\gamma) \le 2^{2t} \|K\|_\infty$$

and

$$\|\tilde{B}_t^{id} K\|_\infty \le \int_\Gamma \|B_t^{id} K\|_\infty d\lambda(\gamma) \le 2t 2^{2(t-1)} \|K\|_\infty + 2^{2(t-1)} \|K\|_\infty$$

holds. Thus the operators $\tilde{B}_t$ and $\tilde{B}_t^{id}$ are bounded and hence continuous, implying the existence of $\tilde{A}_t$ and $\tilde{A}_t^{id}$.

To prove reformulation (5.11) we need some technical facts, that are likely to be well-known to experts in the field. However, we were not able to find proper references apart from an article by Cimprič, Kuhlmann, and Scheiderer [14]. They do not exactly prove the lemmas below but the proofs we give here are just the adjusted versions of proofs they provide in Section 6 in their paper. For a continuous map $f : X \to Y$ and a Borel measure $\mu$ on $X$ we denote the *push forward measure* $\mu(f^{-1}(Z))$ on $Y$ by $f_*(\mu)$.

**Lemma 5.3.4.** *Let $\sigma : \Gamma \times I_t \to I_t$ be the group action defined by $\sigma(\gamma, S) = \gamma S$. A Borel measure $\mu$ on $I_t$ is $\Gamma$-invariant if and only if*

$$\mu = \sigma_*(\lambda \otimes \mu),$$

*where $\lambda$ is the normalized Haar measure on $\Gamma$.*

*Proof.* For $\gamma' \in \Gamma$ and $S \subseteq I_t$ we have

$$\mu(\gamma' S) = \sigma_*(\lambda \otimes \mu)(\gamma' S) = \lambda \otimes \mu(\sigma^{-1}(\gamma' S))$$
$$= \lambda \otimes \mu((\gamma, S) \in \Gamma \times I_t : \gamma S \in \gamma' S)$$
$$\overset{Fubini}{=} \int_\Gamma \mu(S \in I_t : S \in \gamma^{-1} \gamma' S) d\lambda(\gamma)$$
$$\overset{Haar}{=} \int_\Gamma \mu(S \in I_t : S \in S) d\lambda(\gamma) = \int_\Gamma \mu(S) d\lambda(\gamma) = \mu(S).$$

Conversely, if $\mu$ is $\Gamma$-invariant, then for every Borel set $S \subseteq I_t$ we have

$$\sigma_*(\lambda \otimes \mu)(S) = \lambda \otimes \mu(\sigma^{-1}(S)) = \int_{\gamma \in \Gamma} \mu(\gamma^{-1} S) d\lambda = \int_{\gamma \in \Gamma} \mu(S) d\lambda = \mu(S)$$

by the Fubini formula. $\qquad\square$

**Lemma 5.3.5.** *Let $v$ be any Borel measure on $I_t/\Gamma$ and $\pi : I_t \to I_t/\Gamma$ be the quotient map.*

*a) There exists a unique $\Gamma$-invariant measure $\mu \in \mathcal{M}(I_t)$ with $\pi_*(\mu) = v$. We will denote it by $\pi^*(v) := \mu$.*

*b) Explicitly, if $f : I_t \to \mathbb{R}$ is a measurable function, then*

$$\int_{I_t} f(S)d\mu(S) = \int_{I_t/\Gamma} \bar{h}_f(Q)dv(Q),$$

*where the function $\bar{h}_f : I_t/\Gamma \to \mathbb{R} \cup \{\infty\}$ is defined by*

$$\bar{h}_f(\pi(S)) := \int_{\Gamma} f(\gamma S)d\lambda(\gamma) \qquad (S \in I_t).$$

*Proof.* We prove the statements in two parts.

*a) Existence:* We consider again the normalized Haar measure $\lambda$ on $\Gamma$. Let $S \subseteq I_t$ be arbitrary Borel set, $S \in I_t$ a fixed point configuration and the orbit map $o_S : \Gamma \to I_t$ defined by $o_S(\gamma) := \gamma S$. We further define the function

$$h_S(S) := (o_{S*}\lambda)(S) = \lambda\{\gamma \in \Gamma : \gamma S \in S\}.$$

As a function of $\gamma$ onto the Haar measure $\lambda$ of the $\gamma$-cut of the measure spaces $(\Gamma, \mathcal{B}(\Gamma), \lambda)$ and $(I_t, \mathcal{B}(I_t), \mu')$, where $\mu'$ is any $\sigma$-finite Borel measure on $I_t$, the function $h_S$ is measurable (see Lemma 23.2 in [9]). Furthermore, for every $S \in I_t$ it is $\Gamma$-invariant, due to the invariance of the Haar measure $\lambda$:

$$h_S(\gamma_0 S) = \lambda(\{\gamma \in \Gamma : \gamma\gamma_0 S \in S\}) = \lambda\left(\{\gamma\gamma_0^{-1} \in \Gamma : \gamma S \in S\}\right)$$
$$= \lambda(\{\gamma \in \gamma_0\Gamma : \gamma S \in S\}) = \lambda(\{\gamma \in \Gamma : \gamma S \in S\}) = h_S(S).$$

We observe that the fibres of $\pi : I_t \to I_t/\Gamma$ are the $\Gamma$-orbits:

$$\pi^{-1}(\pi(T)) = \{S \in I_t : \pi(S) = \pi(T)\} = \{S \in I_t : \exists\gamma \in \Gamma : S = \gamma T\}$$

and thus $h_S$ induces a measurable function $\bar{h}_S : I_t/\Gamma \to [0, 1]$ by $\bar{h}_S(\pi(S)) = h_S(S)$ for every $S \in I_t$. Now we consider again our Borel measure $v$ on $I_t/\Gamma$ and define

$$\mu(S) := \int_{I_t/\Gamma} \bar{h}_S(Q)dv(Q)$$

for every Borel set $S \subseteq I_t$. Then $\mu$ is a Radon measure on $I_t$, since $(I_t, d'_H)$ is a compact metric space, therefore separable and complete, and thus a Radon space.
For $\Gamma$-invariance we consider $S = \bigcup_{n\in\mathbb{N}} S_n$ as a countable union of pairwise disjoint Borel

sets in $I_t$. Thus we have $h_S(S) = \sum_{n\in\mathbb{N}} h_{S_n}(S)$ defined pointwise on every $S \in I_t$ and therefore $\mu(S) = \sum_{n\in\mathbb{N}} \mu(S_n)$. Additionally, we proved above that $h_{\gamma S} = h_S$ for every $\gamma \in \Gamma$ implying the $\Gamma$-invariance of $\mu$ and furthermore we have $\pi_*(\mu) = \nu$ because

$$\pi_*(\mu)(Q') = \mu(\pi^{-1}(Q')) = \int_{I_t/\Gamma} \bar{h}_{\pi^{-1}(Q')}(Q)d\nu(Q)$$

$$\overset{\text{Fix } P\in\pi^{-1}(Q)}{=} \int_{I_t/\Gamma} h_{\pi^{-1}(Q')}(P)d\nu(Q)$$

$$= \int_{I_t/\Gamma} \lambda(\{\gamma \in \Gamma : \gamma P \in \pi^{-1}(Q')\})d\nu(Q)$$

$$= \int_{I_t/\Gamma} \lambda(\Gamma)\mathbb{1}_{Q\in Q'}d\nu(Q) = \nu(Q').$$

*Uniqueness and b):* Let $f : I_t \to \mathbb{R}_{\geq 0}$ be measurable. The function $h_f : I_t \to \mathbb{R} \cup \{\infty\}$, $h_f(S) := \int_\Gamma f(\gamma S)d\lambda(\gamma)$ is again measurable and $\Gamma$-invariant, so it induces a measurable function $\bar{h}_f : I_t \to \mathbb{R} \cup \{\infty\}$ by $\bar{h}_f(\pi(S)) := \int_\Gamma f(\gamma S)d\lambda(\gamma)$. Given any $\Gamma$-invariant measure $\tilde{\mu}$ on $I_t$ with $\pi_*(\tilde{\mu}) = \nu$, we have

$$\int_{I_t} f(S)d\tilde{\mu}(S) \overset{\text{Lemma 5.3.4 \& Fubini}}{=} \int_{I_t}\int_\Gamma f(\gamma S)d\lambda(\gamma)d\tilde{\mu}(S)$$

$$= \int_{I_t} \bar{h}_f(\pi(S))d\tilde{\mu}(S)$$

$$= \int_{I_t/\Gamma} \bar{h}_f(Q)d\pi_*(\tilde{\mu})(Q)$$

$$= \int_{I_t/\Gamma} \bar{h}_f(Q)d\nu(Q).$$

This establishes on the one hand the uniqueness of $\mu$ by considering $f = \mathbb{1}_S$ and on the other hand part b) of our claim.                                                    □

We have now the necessary ingredients to finally prove Theorem 5.3.3. The proof relies on the fact that for every feasible measure $\mu$ for formulation (5.9) the measure $\nu = \pi_*(\mu)$ is feasible for (5.11) with the same objective value and vice versa with $\nu$ being feasible for (5.11) and consequently $\mu = \pi^*(\nu)$ being feasible for (5.9).

*Proof of Theorem 5.3.3.* Let $\mu \in \mathcal{M}(I_{2t})_{\geq 0}$ be feasible for (5.9), then we define $\nu \in \mathcal{M}(I_{2t}/\Gamma)_{\geq 0}$ by $\nu = \pi_*(\mu)$. We show that $\nu$ is feasible for the program (5.11). For this we have

$\nu(\{\emptyset\}) = \mu(\pi^{-1}\{\emptyset\}) = \mu(\{\emptyset\})$ and for an arbitrary kernel $K \in C(I_t \times I_t)_{\geq 0}$:

$$\langle \tilde{A}_t \nu, K \rangle = \langle \nu, \tilde{B}_t K \rangle = \int_{I_{2t}/\Gamma} \tilde{B}_t K(Q) d\nu(Q)$$

$$\overset{\text{Lemma 5.3.4}}{=} \int_{I_{2t}} B_t K(S) d\mu(S)$$

$$= \langle \mu, B_t K \rangle = \langle A_t \mu, K \rangle \geq 0$$

and similarly

$$\langle \tilde{A}_t^{id} \nu, K \rangle = \langle \nu, \tilde{B}_t^{id} K \rangle = \int_{I_{2t}/\Gamma} \tilde{B}_t^{id} K(Q) d\nu(Q)$$

$$\overset{\text{Lemma 5.3.4}}{=} \int_{I_{2t}} B_t^{id} K(S) d\mu(S)$$

$$= \langle \mu, B_t^{id} K \rangle = \langle A_t^{id} \mu, K \rangle \geq 0.$$

We observe further that

$$\nu(I_{=1}/\Gamma) = \mu(\pi^{-1}(I_{=1}/\Gamma)) = \mu(\{S \in I_t : \pi(S) \in I_{=1}/\Gamma\}) = \mu(I_{=1}),$$

and thus the objective values coincide.
For the opposite inclusion we consider $\nu \in M(I_{2t}/\Gamma)_{\geq 0}$ feasible for (5.11) and define $\mu \in M(I_{2t})_{\geq 0}$ by

$$\mu := \pi^*(\nu).$$

In particular this implies that $\nu = \pi_*(\mu)$ and thus the above arguments still hold in the same way.                                                                                      □

## 5.4    The dual hierarchy

In the preceding sections we achieved a simplified reformulation (5.11) of the covering number by exploiting symmetry inherent to the problem. In the upcoming section we develop a duality theory for these symmetric covering numbers and prove strong duality.

### 5.4.1    Definition of the hierarchy and volume bound

The following lemmas ensure that the functions $\mathbb{1}_{I_{=1}}$ and $\mathbb{1}_{\{\emptyset\}}$ are continuous in $I_t$, which is essential for our approach to a duality theory, since we work with the dual spaces of continuous functions $C(I_t)$ and signed Radon measures $M(I_t)$. We recall the set $(U_1, \ldots, U_k)_t$ (see (5.4)) as environments of a point configuration $Y \in X_t$. The first lemma has also been proven in [23] in the context of topological spaces but to omit the additional notation and definitions we reprove it in the context of metric spaces.

**Lemma 5.4.1.** *[23] The map $I_t \to \mathbb{N}$, $S \mapsto |S|$ is continuous for every $t \in \mathbb{N}$. In particular, $I_{=t}$ is both open and closed.*

*Proof.* Let $\{P_N\}$ be a sequence in $I_t$ converging to an arbitrary subset $\{y_1, \ldots, y_k\} \in I_t$, defined by pairwise different elements $y_i$. By the Hausdorff property of $X$ there exist pairwise disjoint open neighbourhoods $U_i \subseteq B(y_i, \delta)$ for any sufficiently small $\delta > 0$, where $y_i \in U_i$. The set $(U_1, \ldots, U_k)_t$ is open and contains $\{y_1, \ldots, y_k\}$. Hence, we eventually have $P_N \in (U_1, \ldots, U_k)_t$. Then $|P_N| \geq k$ since the $U_i$ are pairwise disjoint and $|P_N| \leq k$ since for every $\{x, x'\} \subseteq P_N \cap U_i$ we have $\varepsilon \leq d(x, x') \leq 2\delta$ leading to a contradiction for sufficiently small $\delta > 0$. $\square$

**Corollary 5.4.2.** *The function $\mathbb{1}_{\{\emptyset\}} : I_t \to \mathbb{N}$ is continuous.*

*Proof.* By Lemma 5.4.1 we know that the sets $I_{=s}$ are both open and closed. We consider the preimage of the function, considering the open sets $\{0\}, \{1\}$, i.e.,

$$\mathbb{1}_{\{\emptyset\}}^{-1}(\{0\}) = \bigcup_{s=1}^{t} I_{=s},$$

which is open as a union of open sets, and

$$\mathbb{1}_{\{\emptyset\}}^{-1}(\{1\}) = \{\emptyset\} = \bigcap_{s=1}^{t}(I_t \setminus I_{=s}),$$

which is open as a finite intersection of open sets. $\square$

We finally show the continuity of $\mathbb{1}_{I_{=1}/\Gamma}$ and $\mathbb{1}_{I_{=0}/\Gamma} = \mathbb{1}_{\{\pi(\emptyset)\}}$. A function $f : I_t/\Gamma \to \mathbb{N}$ is continuous if and only if $f \circ \pi : I_t \to \mathbb{N}$ is continuous (see, e.g., [46]). We apply this to the identity $\mathbb{1}_{I_{=t}/\Gamma}(\pi(S)) = \mathbb{1}_{I_{=t}}(S)$ and obtain the continuity of $\mathbb{1}_{I_{=t}/\Gamma}$, a slightly more general result.

With the help of these technicalities we can apply the theory of general conic programming (see, e.g., Chapter IV, Section 6 in [8]) to state the dual program to (5.11) as

$$\mathcal{N}^t(X, r)^* := \sup y$$
$$y \in \mathbb{R}, \ K \in C(I_t \times I_t)_{\geq 0}, \ K' \in C(I_{t-1} \times I_{t-1})_{\geq 0}$$
$$\mathbb{1}_{I_{=1}/\Gamma}(Q) - y \mathbb{1}_{\{\pi(\emptyset)\}}(Q)$$
$$- \tilde{B}_t K(Q) - \tilde{B}_t^{id} K'(Q) \geq 0 \text{ for all } Q \in I_{2t}/\Gamma$$

and furthermore, by distinguishing the cases $Q = \pi(\emptyset)$ and $Q \neq \pi(\emptyset)$, we obtain

$$\mathcal{N}^t(X, r)^* = \sup y$$
$$y \in \mathbb{R}, \ K \in C(I_t \times I_t)_{\geq 0}, \ K' \in C(I_{t-1} \times I_{t-1})_{\geq 0}$$
$$- y - \tilde{B}_t K(\pi(\emptyset)) - \tilde{B}_t^{id} K'(\pi(\emptyset)) \geq 0$$
$$\mathbb{1}_{I_{=1}/\Gamma}(Q) - \tilde{B}_t K(Q) - \tilde{B}_t^{id} K'(Q) \geq 0 \text{ for all } Q \in I_{2t}/\Gamma \setminus \{\pi(\emptyset)\}.$$

Thus, without loss of generality, we can set $y = -B_t K(\pi(\emptyset)) - B_t^{id} K'(\pi(\emptyset))$ and obtain

$$N^t(X,r)^* = \sup \, - \tilde{B}_t K(\pi(\emptyset)) - \tilde{B}_t^{id} K'(\pi(\emptyset)) \tag{5.12}$$

$$K \in C(I_t \times I_t)_{\geq 0}, \; K' \in C(I_{t-1} \times I_{t-1})_{\geq 0},$$

$$\mathbb{1}_{I_{=1}/\Gamma}(Q) - \tilde{B}_t K(Q) - \tilde{B}_t^{id} K'(Q) \geq 0 \text{ for all } Q \in I_{2t}/\Gamma \setminus \{\pi(\emptyset)\}.$$

This last formulation is especially useful to prove that $N^1(X,r)$ is at least the volume bound $\frac{1}{\omega_r}$.

**Theorem 5.4.3.** *Let $X$ be equipped with a probability measure $\omega$ satisfying properties (a) and (b) in Section 5.1, then*

$$N^1(X,r) = \frac{1}{\omega_r}.$$

*Proof.* From (5.8) we obtain $N^1(X,r) \leq \frac{1}{\omega_r}$. For the reverse inequality we consider $t = 1$. Then we have $\tilde{B}_1 K(\pi(\emptyset)) = K(\emptyset, \emptyset)$, $\tilde{B}_1^{id} K'(\pi(\emptyset)) = -K'(\emptyset, \emptyset)$ and

$$\tilde{B}_1^{id} K' \left( \pi(\{y\}) \right) = \int_\Gamma \sum_{x \in B_r : \{x\} = \gamma\{y\}} K'(\emptyset, \emptyset) d\lambda(\gamma)$$

$$= K'(\emptyset, \emptyset) \int_\Gamma \mathbb{1}_{\gamma\{y\} \in B_r} d\lambda(\gamma)$$

$$= K'(\emptyset, \emptyset) \lambda \left( \{\gamma \in \Gamma : \gamma\{y\} \in B_r\} \right).$$

This leads to the program

$$\sup \, - K(\emptyset, \emptyset) + K'(\emptyset, \emptyset)$$

$$K \in C(I_1 \times I_1)_{\geq 0}, \; K' \in C(I_0 \times I_0)_{\geq 0},$$

$$1 - \tilde{B}_1 K \left( \pi(\{y\}) \right) - \lambda \left( \{\gamma \in \Gamma : \gamma\{y\} \in B_r\} \right) K'(\emptyset, \emptyset) \geq 0 \text{ for every } y \in X$$

We observe that $K = 0$, $K'(\emptyset, \emptyset) = \frac{1}{\lambda(\{\gamma \in \Gamma : \gamma\{y\} \in B_r\})}$ is a feasible solution for this program with objective value $\frac{1}{\lambda(\{\gamma \in \Gamma : \gamma\{y\} \in B_r\})} = \frac{\lambda(\Gamma)}{\lambda(\{\gamma \in \Gamma : \gamma\{y\} \in B_r\})}$. Since the metric space is equipped with a probability measure $\omega$ satisfying properties (a) and (b) from Section 5.1, we have that

$$\frac{\lambda(\Gamma)}{\lambda(\{\gamma \in \Gamma : \gamma\{y\} \in B_r\})} = \frac{\omega(X)}{\omega(B(x,r))} \leq N^1(X,r)$$

holds due to the transitivity of $\Gamma$.                                                      $\square$

## 5.4.2   Strong duality

We follow an approach by Barvinok (see Chapter IV, Section 7 in [8]) to prove strong duality. For this we need to reformulate the primal program $N^t(X,r)$ in the form

$$\inf\{\langle x, c \rangle : Ax = b, \; x \in \mathcal{K}\}.$$

Therefore we define the spaces $E = M(I_{2t}/\Gamma) \oplus M(I_t \times I_t)_{sym} \oplus M(I_{t-1} \times I_{t-1})_{sym}$ and $F = C(I_{2t}/\Gamma) \oplus C(I_t \times I_t)_{sym} \oplus C(I_{t-1} \times I_{t-1})_{sym}$ with duality

$$\langle e_1 + e_2 + e_3, f_1 + f_2 + f_3 \rangle := \int_{I_{2t}/\Gamma} f_1 de_1 + \int_{I_t \times I_t} f_2 de_2 + \int_{I_{t-1} \times I_{t-1}} f_3 de_3.$$

Then, by choosing $c = (\mathbb{1}_{I_{=1}}, 0, 0) \in F$ and $\hat{A} : E \to M(I_t \times I_t)_{sym} \oplus M(I_{t-1} \times I_{t-1})_{sym}$ defined by

$$\hat{A}(\mu, \nu, \nu^{id}) = \begin{pmatrix} \tilde{A}_t \mu - \nu \\ \tilde{A}_t^{id} \mu - \nu^{id} \end{pmatrix},$$

we have

$$N^t(\varphi) = \inf \langle (\mu, \nu, \nu^{id}), c \rangle \tag{5.13}$$

$$(\mu, \nu, \nu^{id}) \in M(I_{2t}/\Gamma)_{\geq 0} \times M(I_t \times I_t)_{\geq 0} \times M(I_{t-1} \times I_{t-1})_{\geq 0},$$

$$\hat{A}(\mu, \nu, \nu^{id}) = 0,$$

$$\mu(\{\emptyset\}) = 1.$$

We apply now Theorem 7.2 of Barvinok's book [8], that can be stated for our situation as follows.

**Theorem 5.4.4.** *Suppose that the cone*

$$\hat{\mathcal{K}} = \Big\{ \Big( \hat{A}(\mu, \nu, \nu^{id}), \mu(\{\emptyset\}), \mu(I_{=1}) \Big) :$$

$$(\mu, \nu, \nu^{id}) \in M(I_{2t}/\Gamma)_{\geq 0} \times M(I_t \times I_t)_{\geq 0} \times M(I_{t-1} \times I_{t-1})_{\geq 0} \Big\}$$

*is closed in* $M(I_t \times I_t)_{sym} \times M(I_{t-1} \times I_{t-1})_{sym} \times \mathbb{R} \times \mathbb{R}$ *and that there is a primal feasible measure* $(\mu, \nu, \nu^{id})$. *Then* $N^t(\varphi) = N^t(\varphi)^*$.

We already know that $N^t(X, r)$ has a feasible solution $\mu^* = \sum_{Z \in I_{2t} : Z \subseteq Y} \delta_Z$ for any finite covering $Y \subseteq X$ as was shown in Subsection 5.2.3. Thus it remains to show that $\hat{\mathcal{K}}$ is closed. We observe that $\hat{\mathcal{K}}$ is the Minkowski difference of

$$\mathcal{K}_1 = \Big\{ \Big( \tilde{A}_t \mu, \tilde{A}_t^{id} \mu, \mu(\{\emptyset\}), \mu(I_{=1}) \Big) : \mu \in M(I_{2t}/\Gamma)_{\geq 0} \Big\}$$

and

$$\mathcal{K}_2 = \Big\{ \Big( \nu, \nu^{id}, 0, 0 \Big) : \nu \in M(I_t \times I_t)_{\geq 0}, \nu^{id} \in M(I_{t-1} \times I_{t-1})_{\geq 0} \Big\}.$$

By a theorem of Klee [52] and Dieudonné [26] the Minkowski difference $\mathcal{K}_1 - \mathcal{K}_2$ of two cones is closed if the three conditions

(i) $\mathcal{K}_1 \cap \mathcal{K}_2 = \{0\}$,

(ii) $\mathcal{K}_1$ and $\mathcal{K}_2$ are closed,

(iii) $\mathcal{K}_1$ is locally compact

are satisfied. Before we start proving this, we provide a small lemma that is needed for the proofs of condition (i) and the proof that $\mathcal{K}_1$ is closed and locally compact. In general, we follow closely a proof strategy established by de Laat and Vallentin (see Lemmas 6.7.1, 6.7.3 and 6.7.5 in [23]) and incorporate the symmetrization with respect to $\Gamma$.

**Lemma 5.4.5.** *The operator* $\tilde{B}_t : C(I_t \times I_t) \to C(I_{2t}/\Gamma)$ *defined as above by*

$$\tilde{B}_t K(\pi(S)) := \int_\Gamma \sum_{J,J' \in I_t:\ J \cup J' = \gamma S} K(J, J') d\lambda(\gamma)$$

*is surjective.*

*Proof.* We fix a function $g \in C(I_{2t}/\Gamma)$ and observe that $u : I_t \times I_t \to X_{2t}$, $u(J, J') \mapsto J \cup J'$ is continuous (see Prop. 2.14 in Handel [43]). Hence

$$h : u^{-1}(I_{2t}) \to \mathbb{R}, \ h(J, J') = \frac{g(\pi(J \cup J'))}{\tilde{B}_t \mathcal{J}(J \cup J')},$$

where $\mathcal{J}$ is the symmetric kernel evaluating to 1 everywhere, is continuous due to the continuity of $g \circ \pi$. As a preimage of a closed set $I_{2t}$ with respect to a continuous function $u$ we have that $u^{-1}(I_{2t})$ is closed in the compact Hausdorff space $I_t \times I_t$. For normal spaces, including compact Hausdorff spaces, Tietze's extension theorem provides the existence of a function $H \in C(I_t \times I_t)$ such that $H(J, J') = h(J, J')$ for every $J, J' \in u^{-1}(I_{2t})$. Finally, we observe for each $Q \in I_{2t}/\Gamma$ denoted by $\pi(S) = Q$

$$\tilde{B}_t H(\pi(S)) = \int_\Gamma \sum_{J,J' \in I_t:\ J \cup J' = \gamma S} H(J, J') d\lambda(\gamma) = \int_\Gamma \sum_{J,J' \in I_t:\ J \cup J' = \gamma S} h(J, J') d\lambda(\gamma)$$

$$= \int_\Gamma \sum_{J,J' \in I_t:\ J \cup J' = \gamma S} \frac{g(\pi(J \cup J'))}{\tilde{B}_t \mathcal{J}(\pi(J \cup J'))} d\lambda(\gamma)$$

$$= \int_\Gamma \frac{g(\pi(S))}{\tilde{B}_t \mathcal{J}(\pi(S))} |\{J, J' \in I_t :\ J \cup J' = \gamma S\}| d\lambda(\gamma) = g(\pi(S)).$$

$\square$

We now want to verify conditions (i) – (iii) starting with condition (i). Here the proof is basically a slight modification of the one of Lemma 6.7.3 in de Laat [22].

**Theorem 5.4.6.** $\mathcal{K}_1 \cap \mathcal{K}_2 = \{0\}$

*Proof.* We will show that $\mu \in M(I_{2t}/\Gamma)_{\geq 0}$ with $\mu(\{\emptyset\}) = 0$ is the zero measure if $\tilde{A}_t\mu \in M(I_t \times I_t)_{\geq 0}$. For this we consider the kernel $K \in C(I_t \times I_t)_{sym}$ defined by

$$K(J, J') := \begin{cases} 1 \text{ if } J = J' = \emptyset, \\ 0 \text{ otherwise.} \end{cases}$$

Then we observe $\tilde{A}_t\mu(\{(\emptyset, \emptyset)\}) = \langle \tilde{A}_t\mu, K \rangle = \langle \mu, \tilde{B}_t K \rangle = \mu(\{\emptyset\}) = 0$. Furthermore we define a sequence of functions $f_n \in C(I_t)$, where $n \in \mathbb{Z}$ by

$$f_n(S) := \begin{cases} |n| \text{ if } S = \emptyset, \\ \frac{1}{n} \text{ otherwise.} \end{cases}$$

Since $f_n \otimes f_n \in C(I_t \times I_t)_{\geq 0}$ and $\tilde{A}_t\mu \in M(I_t \times I_t)_{\geq 0}$ we have that $\langle \tilde{A}_t\mu, f_n \otimes f_n \rangle \geq 0$. On the other hand $\langle \tilde{A}_t\mu, f_n \otimes f_n \rangle$ equates to

$$n^2 \tilde{A}_t\mu(\{(\emptyset, \emptyset)\}) + \frac{1}{n^2}\tilde{A}_t\mu(I_t \setminus \{\emptyset\} \times I_t \setminus \{\emptyset\}) + 2\text{sign}(n)\tilde{A}_t\mu(\{\emptyset\} \times I_t \setminus \{\emptyset\}).$$

The first term evaluates to zero. Thus we consider the remaining inequality

$$\frac{1}{n^2}\tilde{A}_t\mu(I_t \setminus \{\emptyset\} \times I_t \setminus \{\emptyset\}) + 2\text{sign}(n)\tilde{A}_t\mu(\{\emptyset\} \times I_t \setminus \{\emptyset\}) \geq 0,$$

let $n$ tend to plus or minus infinity and conclude that $\tilde{A}_t\mu(\{\emptyset\} \times I_t \setminus \{\emptyset\}) = 0$. Next we show that $\mu(I_t/\Gamma) = 0$ with the help of the kernel $L$ defined by

$$L(J, J') := \begin{cases} 1 \text{ if } J = J' = \emptyset, \\ \frac{1}{2} \text{ if } J = \emptyset \text{ or } J' = \emptyset, \\ 0 \text{ otherwise} \end{cases}$$

because then we have

$$\mu(I_t/\Gamma) = \int_{I_t/\Gamma} \int_\Gamma 1 d\lambda(\gamma)d\mu(\pi(S)) = \int_{I_t/\Gamma} \int_\Gamma \mathbb{1}_{I_t}(\gamma S)d\lambda(\gamma)d\mu(\pi(S))$$

$$= \int_{I_t/\Gamma} \int_\Gamma \mathbb{1}_{\{\emptyset\}}(\gamma S) + 2 \cdot \frac{1}{2}\mathbb{1}_{I_t \setminus \{\emptyset\}}(\gamma S)d\lambda(\gamma)d\mu(\pi(S))$$

$$= \int_{I_t/\Gamma} \int_\Gamma \sum_{J, J' \in I_t: \ J \cup J' = \gamma S} L(J, J')d\lambda(\gamma)d\mu(\pi(S))$$

$$= \langle \mu, \tilde{B}_t L \rangle = \langle \tilde{A}_t\mu, L \rangle = \tilde{A}_t\mu(\{(\emptyset, \emptyset)\}) + \tilde{A}_t\mu(\{\emptyset\} \times I_t \setminus \{\emptyset\}) = 0.$$

Finally we show $\mu(I_{2t}/\Gamma) = 0$. If $V$ is a sufficiently small open set in $I_t$, then the union of two distinct elements $J, J' \in V$ can have at most $t$ elements because of the fixed minimal distance $\varepsilon > 0$. Thus $J \cup J' \notin I_{2t} \setminus I_t$ implying

$$\tilde{B}_t(\mathbb{1}_V \otimes \mathbb{1}_V)(\pi(S)) = \int_\Gamma \sum_{J, J' \in I_t: J \cup J' = \gamma S} \mathbb{1}_V(J)\mathbb{1}_V(J')d\lambda(\gamma) = 0$$

whenever $S \in I_{2t} \setminus I_t$. Let $J$ and $J'$ be arbitrary elements in $I_t$, and let $U$ and $U'$ be small open neighborhoods around $J$ and $J'$. For $s = \pm 1$ we have

$$
\begin{aligned}
0 &\leq \langle \tilde{A}_t \mu, (\mathbb{1}_U + s\mathbb{1}_{U'}) \otimes (\mathbb{1}_U + s\mathbb{1}_{U'}) \rangle \\
&= \tilde{A}_t \mu(U \times U) + \tilde{A}_t \mu(U' \times U') + 2s\tilde{A}_t \mu(U \times U'),
\end{aligned}
$$

where the inequality follows because Urysohn's Lemma says that $\mathbb{1}_U + s\mathbb{1}_{U'}$ can be approximated arbitrarily well by continuous functions. Since $\tilde{B}_t K$ is bounded and $\mu|_{I_t/\Gamma} = 0$, we have

$$
\tilde{A}_t \mu(U \times U) = \tilde{A}_t \mu(U' \times U') = 0
$$

for $U$ and $U'$ small enough. This shows $2s\tilde{A}_t \mu(U \times U') \geq 0$, and since $s = \pm 1$ we have $\tilde{A}_t \mu(U \times U') = 0$. Since $J$ and $J'$ are arbitrary this shows $\tilde{A}_t \mu = 0$, and since $\tilde{A}_t$ is injective due to Lemma 5.4.5, we have $\mu = 0$. □

For conditions (ii) and (iii) we first observe that $\mathcal{K}_2 = M(I_t \times I_t)_{\geq 0} \times M(I_{t-1} \times I_{t-1})_{\geq 0} \times \{0\} \times \{0\}$ is closed as a direct product of four closed convex cones. For the remaining conditions on $\mathcal{K}_1$ we need a bit of further background: A *convex base* $\mathcal{K}_B$ of a cone $\mathcal{K}$ is defined as a convex subset of the cone such that every non-zero $x \in \mathcal{K}$ is uniquely determined as a positive multiple of an element in $\mathcal{K}_B$. Additionally, we state a theorem of Klee and Dieudonné [52]:

**Lemma 5.4.7.** *[52] A non-empty pointed cone in a locally convex vector space is closed and locally compact if and only if it admits a compact convex base.*

Finally we prove the remaining conditions on $\mathcal{K}_1$.

**Theorem 5.4.8.** *$\mathcal{K}_1$ is closed and locally compact.*

*Proof.* First we show that

$$
\mathcal{K}_B = \{\mu \in M(I_{2t}/\Gamma)_{\geq 0} : \langle \mathbb{1}_{I_{2t}/\Gamma}, \mu \rangle = 1\}
$$

is compact. Consider the map $M(I_{2t}/\Gamma)_{\geq 0} \to \mathbb{R}$, $\mu \mapsto \langle \mathbb{1}_{I_{2t}/\Gamma}, \mu \rangle = \|\mu\|$, that is continuous. Thus we have that its preimage of $\{1\}$ is a closed subset of $\{\mu \in M(I_{2t}/\Gamma) : \|\mu\| \leq 1\}$, which itself is compact by the Banach-Alaoglu Theorem applied on the Banach space $(C(I_{2t}/\Gamma), \|\cdot\|_\infty)$ equipped with the supremum norm.

Consequently as a closed subset of a compact space, $\mathcal{K}_B$ is compact as well. Due to Lemma 5.4.5 $\tilde{A}_t$ is injective and thus the map $\mu \mapsto (\tilde{A}_t \mu, \tilde{A}_t^{id} \mu, \mu(\{\emptyset\}), \mu(S_{=1}))$ is injective. The image of $\mathcal{K}_B$ under this map is now both compact, due to the continuity of the operator and a convex base due to linearity and injectivity of the operator. Finally we can apply Lemma 5.4.7 and finish the proof. □

## 5.5   Finite convergence of the hierarchy

For the next paragraph we observe that if we fix $\varepsilon > 0$, this determines a *packing problem* on the compact metric space $X$, i.e.,

$$\alpha(X, \varepsilon) := \max\{|Y| : Y \subseteq X, \ B(x, \varepsilon) \cap B(y, \varepsilon) = \emptyset \text{ for all } x, y \in Y \text{ such that } x \neq y\}.$$

This finite *packing number* gives a maximum on the cardinality of sets $S \in I = \{S \subseteq X : d(s_1, s_2) > \varepsilon\}$. For the remainder of the chapter we will simply denote the packing number $\alpha(X, \varepsilon)$ by $\alpha$.

Additionally, the following way to sketch the *inclusion-exclusion principle* for finite sets is a crucial tool. Given two finite sets $A$ and $C$ then

$$\sum_{B : A \subseteq B \subseteq C} (-1)^{|B|} = (-1)^{|A|} \sum_{B \subseteq C \setminus A} (-1)^{|B|}$$

$$= (-1)^{|A|} \sum_{i=0}^{|C \setminus A|} \binom{|C \setminus A|}{i} 1^{|C \setminus A| - i} (-1)^i$$

$$= (-1)^{|A|} (1 - 1)^{|C \setminus A|} = \begin{cases} (-1)^{|A|} & \text{if } A = C \\ 0 & \text{otherwise.} \end{cases}$$

Another crucial definition for the next steps is the one of *characteristic measures* $\chi_R$. We define such a measure for a fixed $R \in I/\Gamma$ with fixed representative $R_\pi$ componentwise by

$$\chi_R : I/\Gamma \to \mathbb{R}, \ \chi_R(Q) := \sum_{P \subseteq R_\pi} \delta_{\pi(P)}(Q),$$

where

$$\delta_{\pi(P)}(Q) = \begin{cases} 1 & \text{if } Q = \pi(P) \\ 0 & \text{otherwise.} \end{cases}$$

Furthermore, these measures define for a fixed $f \in C(I/\Gamma)$ a function, $R \mapsto \chi_R(f)$, where

$$\chi_R(f) := \sum_{P \subseteq R_\pi} \delta_{\pi(P)}(f) = \sum_{P \subseteq R_\pi} f(\pi(P)).$$

With these ingredients it is possible to describe every feasible solution for $N^\alpha(X, r)$, the $\alpha$-th step of (5.11), as an "infinite" convex combination of vectors $\chi_R$:

**Lemma 5.5.1.** *Suppose $\mu \in \mathcal{M}(I_{2t})_{\geq 0}$ is feasible for $N^\alpha(X, r)$. Then there exists a unique probability measure*

$$\sigma \in \mathcal{P}(I/\Gamma) = \{\lambda \in \mathcal{M}(I/\Gamma)_{\geq 0} : \|\lambda\| = 1\}$$

*such that $\mu = \int \chi_R d\sigma(R)$.*

*Proof.* Similar to de Laat and Vallentin [23] we split the proof in four parts.

*Existence:* We denote an arbitrary but fixed representative of a set $Q \in I/\Gamma$ by $Q_\pi$, i.e., $\pi(Q_\pi) = Q$. Let us consider

$$\sum_{R \subseteq Q_\pi} (-1)^{|Q_\pi \setminus R|} \chi_{\pi(R)} = \sum_{R \subseteq Q_\pi} (-1)^{|Q_\pi \setminus R|} \sum_{P \subseteq R} \delta_{\pi(P)}$$

$$= \sum_{P \subseteq Q_\pi} \delta_{\pi(P)} \sum_{R: P \subseteq R \subseteq Q_\pi} (-1)^{|Q_\pi \setminus R|} = \delta_{\pi(Q_\pi)} = \delta_Q,$$

where the second last equality holds due to the inclusion-exclusion principle. Consequently by using the theory of weak vector integrals (see, e.g., [32]) we obtain

$$\mu = \int \delta_Q d\mu(Q) = \int \sum_{R \subseteq Q_\pi} (-1)^{|Q_\pi \setminus R|} \chi_{\pi(R)} d\mu(Q).$$

For a function $f \in C(I/\Gamma)$ we define the linear map

$$l : C(I/\Gamma) \to \mathbb{R}, \ f \mapsto \int \sum_{R \subseteq Q_\pi} (-1)^{|Q_\pi \setminus R|} f(\pi(R)) d\mu(Q)$$

and observe that since $\mu$ is nonnegative, its image $\text{Im}(l) \subseteq \mathbb{R}$ can be bounded by

$$\left| \int \sum_{R \subseteq Q_\pi} (-1)^{|Q_\pi \setminus R|} f(\pi(R)) d\mu(Q) \right| \leq \int \left| \sum_{R \subseteq Q_\pi} (-1)^{|Q_\pi \setminus R|} \|f\|_\infty \right| d\mu(Q)$$

$$\leq \int 2^{|Q_\pi|} \|f\|_\infty d\mu(Q)$$

$$\leq \int 2^\alpha \|f\|_\infty d\mu(Q)$$

$$\leq 2^\alpha \|f\|_\infty \|\mu\|.$$

Thus we have that $l$ is bounded and thus defines a signed Radon measure $\sigma$ on $I/\Gamma$ by applying the Riesz-Representation Theorem on $l$ and $\chi_R(f) \in C(I/\Gamma)$:

$$\int \chi_R(f) d\sigma(R) \overset{Riesz}{=} l(\chi_R(f)) = \int \sum_{R \subseteq Q_\pi} (-1)^{|Q_\pi \setminus R|} \chi_{\pi(R)}(f) d\mu(Q) = \langle f, \mu \rangle$$

for each $f \in C(I/\Gamma)$, so $\mu = \int \chi_R d\sigma(R)$.

*Uniqueness:* If $\sigma' \in \mathcal{M}(I_{2t}/\Gamma)$ is another measure such that $\mu = \int \chi_R d\sigma'(R)$, then $\int \chi_R d(\sigma - \sigma')(R) = 0$. If we evaluate this at a Borel set $L \subseteq I_{=t}/\Gamma$ with $t = \alpha$ we obtain

$$0 = \int_{I_{2t}/\Gamma} \chi_R(L) d(\sigma - \sigma')(R) = \int_{I_{2t}/\Gamma} \delta_R(L) d(\sigma - \sigma')(R) = (\sigma - \sigma')(L).$$

Therefore we conclude $\sigma\big|_{I_{=t}/\Gamma} = \sigma'\big|_{I_{=t}/\Gamma}$. Furthermore if we repeat the argumentation for $t = \alpha - 1, \ldots, 1, 0$, we obtain $\sigma = \sigma'$ implying uniqueness of $\sigma$.

*Positivity:* Let $g \in C(I/\Gamma)_{\geq 0}$ be arbitrary and define $f \in C(I)$ by

$$f(S) = \sum_{P \subseteq S} (-1)^{|S \setminus P|} \sqrt{g(\pi(P))}$$

and conclude

$$\sum_{S \subseteq R} f(S) = \sum_{S \subseteq R} \sum_{P \subseteq S} (-1)^{|S \setminus P|} \sqrt{g(\pi(P))} = \sum_{P \subseteq R} (-1)^{|P|} \sqrt{g(\pi(P))} \sum_{S : P \subseteq S \subseteq R} (-1)^{|S|}$$

$$= \sum_{P = R} (-1)^{|P|} \sqrt{g(\pi(P))} (-1)^{|P|} = \sqrt{g(\pi(R))}.$$

We have

$$0 \leq \langle f \otimes f, \tilde{A}_\alpha \mu \rangle = \langle \tilde{B}_\alpha(f \otimes f), \mu \rangle = \int \chi_R (\tilde{B}_\alpha(f \otimes f)) d\sigma(R)$$

$$= \int \sum_{P \subseteq R_\pi} \delta_{\pi(P)} (\tilde{B}_\alpha(f \otimes f)) d\sigma(R) = \int \sum_{P \subseteq R_\pi} \tilde{B}_\alpha(f \otimes f)(\pi(P)) d\sigma(R)$$

$$= \int \sum_{P \subseteq R_\pi} \int_\Gamma \sum_{J \cup J' = \gamma P} f(J) f(J') d\lambda(\gamma) d\sigma(R)$$

$$= \int \int_\Gamma \sum_{J \cup J' \subseteq \gamma R_\pi} f(J) f(J') d\lambda(\gamma) d\sigma(R)$$

$$= \int \int_\Gamma \left( \sum_{P \subseteq \gamma R_\pi} f(P) \right)^2 d\lambda(\gamma) d\sigma(R)$$

$$= \int \int_\Gamma g(\pi(\gamma R_\pi)) d\lambda(\gamma) d\sigma(R) = \int \int_\Gamma g(R) d\lambda(\gamma) d\sigma(R) = \int g(R) d\sigma(R)$$

implying that $\langle g, \sigma \rangle \geq 0$ for every $g \in C(I/\Gamma)_{\geq 0}$. Thus $\sigma$ is a positive measure.

*Normalization:* Due to the feasibility of $\mu$ for $\mathcal{N}^\alpha(X, r)$ we have

$$\|\sigma\| = \int d\sigma(S) = \int \chi_S (\pi(\{\emptyset\})) d\sigma(S) = \mu(\pi(\{\emptyset\})) = 1$$

implying that $\sigma$ is a probability measure.                                                    $\square$

We will now argue that the support of every feasible measure $\mu$ for the last step of our hierarchy $\mathcal{N}^\alpha(X, r)$ is contained in the subsets of $X$ that define a covering of $X$. For this we will show that for every non-covering set, we can find a positive semidefinite kernel $K \in C(I_\alpha \times I_\alpha)_{\geq 0}$ such that $\langle \tilde{A}_\alpha^{id} \mu, K \rangle < 0$ and thus the covering constraint is not satisfied.

**Lemma 5.5.2.** *For a non-empty ball $B_r \subseteq X$ let $k : I_\alpha \to \mathbb{R}$,*

$$k(J) := \begin{cases} (-1)^{|J|} \text{ if } J \subseteq B_r \\ 0 \text{ otherwise,} \end{cases}$$

*then there exists a continuous function $g : I_\alpha \to \mathbb{R}$ such that $k(J) = g(J)$ $v$-almost surely for every $v \in \mathcal{M}(I_\alpha)$.*

*Proof.* First we observe that

$$k(J) = \left( \mathbb{1}_{2\left\lfloor \frac{1}{2}(|J|) \right\rfloor} - \mathbb{1}_{2\left\lfloor \frac{1}{2}(|J|+1) \right\rfloor} \right) \mathbb{1}_{J \subseteq B_r}$$

$$= \left( \sum_{t=0,\ t \text{ even}}^{|J|} \mathbb{1}_{I_{=t}} - \sum_{t=0,\ t \text{ odd}}^{|J|} \mathbb{1}_{I_{=t}} \right) \mathbb{1}_{J \subseteq B_r}.$$

Due to the continuity of $\mathbb{1}_{I_{=t}}$, given by Lemma 5.4.1, it suffices to show that there is a continuous function $g$ such that $g(J) = \mathbb{1}_{J \subseteq B_r}$ $v$-almost surely.

We further consider $\{J \in I_\alpha : J \not\subseteq B_r\}$ and show that it is an open set in $I_\alpha$. Because $J \not\subseteq B_r$, implying $J \neq \emptyset$, there is an element $y_0 \in J \setminus B_r$. Due to the Hausdorff property of $X$ the minimal distance we have $d(y_0, c) > \delta$ for every $c \in B_r$. Thus we can consider the ball $B(J, \delta)$ defined by the Hausdorff metric and show that for nonempty $B_r$ $d_H(B_r, B(J, \delta)) \geq \inf_{c \in B_r} d(c, B(J, \delta)) > 0$ holds. This implies that

$$A := \{J \in I_\alpha : J \not\subseteq B_r\} \text{ is open and } B := \{J \in I_\alpha : J \subseteq B_r\} \text{ is closed.}$$

As a metric space $(I_\alpha, d_H')$ is a normal space and thus we can apply Urysohn's Lemma, which gives us that for any compact subset of $A$, say $A'$ and $B$ there exists a continuous function $g$ with $g(a) = 0$ for every $a \in A'$ and $g(b) = 1$ for every $b \in B$. In the case of our metric space $(I_\alpha, d_H')$ one $g$ that satisfies the constraints can be shown to be $g(J) = \frac{d_H'(J, A')}{d_H'(J, A') + d_H'(J, B)}$. This $g$ is bounded by 0 and 1. For an arbitrary Radon measure $v \in \mathcal{M}(I_\alpha)$ and every compact $A' \subseteq A$ we consider

$$\int_{I_\alpha} |\mathbb{1}_{J \subseteq B_r} - g| dv = \int_{A \setminus A'} |\mathbb{1}_{J \subseteq B_r} - g| dv \leq v(A \setminus A').$$

Thus

$$\int_{I_\alpha} |\mathbb{1}_{J \subseteq B_r} - g| dv \leq \inf_{A' \subseteq A,\ A' \text{ compact}} v(A \setminus A') \leq v(A) - \sup_{A' \subseteq A,\ A' \text{ compact}} v(A') = 0,$$

where the last step is due to the inner regularity of every Radon measure $v$.                  □

**Theorem 5.5.3.** *Suppose $\mu$ is feasible for $N^\alpha(X, r)$. Then the objective value $\mu(I_{=1})$ upper bounds the covering number $N(X, r)$.*

*Proof.* We consider a feasible measure $\mu$ for the program $\mathcal{N}^\alpha(X, r)$ and its representation $\mu = \int \chi_R d\sigma(R)$, which exists due to Lemma 5.5.1. We observe that $\mu$ satisfies the covering constraint, i.e., for every $K \in C(I_\alpha \times I_\alpha)_{\geq 0}$ we have:

$$0 \leq \langle \tilde{A}_\alpha^{id} \mu, K \rangle = \langle \mu, \tilde{B}_\alpha^{id} K \rangle = \int \chi_R(\tilde{B}_\alpha^{id} K) d\sigma(R) = \int \sum_{P \subseteq R_\pi} \tilde{B}_\alpha^{id}(\pi(P)) d\sigma(R)$$

$$= \int \sum_{P \subseteq R_\pi} \int_\Gamma \sum_{x \in B_r, \ J,J' \in I_\alpha: \ J \cup J' \cup \{x\} = \gamma P} K(J, J') - \sum_{J,J' \in I_\alpha: \ J \cup J' = \gamma P} K(J, J') d\lambda(\gamma) d\sigma(R)$$

$$= \int \int_\Gamma \sum_{x \in B_r, \ J,J' \in I_\alpha: \ J \cup J' \cup \{x\} \subseteq \gamma R_\pi} K(J, J') - \sum_{J,J' \in I_\alpha: \ J \cup J' \subseteq \gamma R_\pi} K(J, J') d\lambda(\gamma) d\sigma(R)$$

$$= \int \int_\Gamma \sum_{x \in \gamma R_\pi \cap B_r, \ J,J' \subseteq \gamma R_\pi} K(J, J') - \sum_{J,J' \subseteq \gamma R_\pi} K(J, J') d\lambda(\gamma) d\sigma(R)$$

$$= \int \int_\Gamma |\gamma R_\pi \cap B_r| \sum_{J,J' \subseteq \gamma R_\pi} K(J, J') - \sum_{J,J' \subseteq \gamma R_\pi} K(J, J') d\lambda(\gamma) d\sigma(R).$$

In particular this holds for the continuous kernel $g \otimes g \in C(I_\alpha \times I_\alpha)_{\geq 0}$ defined in Lemma 5.5.2, which equals the following kernel $\mu$-almost surely:

For a ball $B_r \subseteq X$ we define $k : I_\alpha \to \mathbb{R}$, $k(J) := \begin{cases} (-1)^{|J|} & \text{if } J \subseteq B_r \\ 0 & \text{otherwise} \end{cases}$ and furthermore

$$\sum_{J,J' \in I_\alpha: J, J' \subseteq \gamma R_\pi} k \otimes k(J, J') = \sum_{J,J' \in I_\alpha: J,J' \subseteq \gamma R_\pi \cap B_r} (-1)^{|J|}(-1)^{|J'|}$$

$$= \left( \sum_{J \in I_\alpha: J \subseteq \gamma R_\pi \cap B_r} (-1)^{|J|} \right)^2 = \mathbb{1}_{\gamma R_\pi \cap B_r = \emptyset},$$

where the last step is due to the inclusion-exclusion principle mentioned above. Applying this to the covering constraint yields

$$0 \leq \int \int_\Gamma |\gamma R_\pi \cap B_r| \sum_{J,J' \subseteq \gamma R_\pi} g \otimes g(J, J') - \sum_{J,J' \subseteq \gamma R_\pi} g \otimes g(J, J') d\lambda(\gamma) d\sigma(R)$$

$$= \int \int_\Gamma |\gamma R_\pi \cap B_r| \sum_{J,J' \subseteq \gamma R_\pi} k \otimes k(J, J') - \sum_{J,J' \subseteq \gamma R_\pi} k \otimes k(J, J') d\lambda(\gamma) d\sigma(R)$$

$$= \int \int_\Gamma (|\gamma R_\pi \cap B_r| - 1) \mathbb{1}_{\gamma R_\pi \cap B_r = \emptyset} d\lambda(\gamma) d\sigma(R)$$

$$= \int -\lambda(\gamma \in \Gamma : \ \gamma R_\pi \cap B_r = \emptyset) d\sigma(R)$$

If we assume that $R_\pi$ is not a covering, we have that $\lambda(\gamma \in \Gamma : \gamma R_\pi \cap B_r = \emptyset)$ is strictly positive, since $\Gamma$ is transitive, a contradiction. Thus we have shown that any feasible measure for $\mathcal{N}^\alpha(X, r)$ is only supported on coverings of $X$. We finish the proof by looking at the objective function of $\mathcal{N}^\alpha(X, r)$

$$\mu(I_{=1}/\Gamma) = \int \chi_R(I_{=1}/\Gamma)d\sigma(R) = \int |R_\pi|d\sigma(R) \geq \int \mathcal{N}(X, r)d\sigma(R) = \mathcal{N}(X, r).$$

The reverse implication follows by definition of $\mathcal{N}^t(X, r)$.                                    □

## 5.6   Symmetry-reduced dual hierarchy

We have now completed a number of theoretic results on the covering number $\mathcal{N}(X, r)$. The main obstacle for computations of such a geometric problem is the size of the sample that is needed to discretize the problem

$$\mathcal{N}^t(X, r) = \sup \ - \tilde{B}_t K(\pi(\emptyset)) - \tilde{B}_t^{id} K'(\pi(\emptyset)) \tag{5.14}$$
$$K \in C(I_t \times I_t)_{\geq 0}, \ K' \in C(I_{t-1} \times I_{t-1})_{\geq 0},$$
$$\mathbb{1}_{I_{=1}/\Gamma}(Q) - \tilde{B}_t K(Q) - \tilde{B}_t^{id} K'(Q) \geq 0 \text{ for all } Q \in I_{2t}/\Gamma \setminus \{\pi(\emptyset)\}. \tag{5.15}$$

appropriately. Although we already used symmetries with respect to $\Gamma$ to reduce the number of constraints in the dual program (5.12) above, we still need to sample the kernels $K$ and $K'$ and the constraints $Q \in I_{2t}/\Gamma \setminus \{\pi(\emptyset)\}$, which is very expensive in terms of computation time.

In the related problems of *energy minimization* (see [21]) and *packing of superballs* (see [28]) a fruitful approach to reduce the number of necessary sample points is to exploit symmetries inherent in the problem.

Therefore, in the remainder of this section, we reduce the degrees of freedom of the kernels $K$ and $K'$ by exploiting symmetries to achieve a reformulation of (5.14) that is computable for certain choices of $X$. For this we consider $B_r = B(e, r)$, i.e., we denote the center of $B_r$ by $e$. The *stabilizer subgroup* $H \subseteq \Gamma$ with respect to the point $e$ is defined by

$$H := \{\gamma \in \Gamma : \gamma e = e \text{ for every } x \in X\}.$$

As a subset of isometries the group $H$ in particular satisfies $\gamma B(e, r) = B(e, r)$ for all $\gamma \in H$. Additionally, we say that a kernel $K \in C(I_t \times I_t)_{\geq 0}$ is $\Gamma$- respectively $H$-invariant if and only if the shifted kernel $K^\gamma$ defined by $K^\gamma(J, J') = K(\gamma J, \gamma J')$ and $K$ coincide, i.e.,

$$K \in C(I_t \times I_t)_{\geq 0}^\Gamma \iff K(J, J') = K(\gamma J, \gamma J') \text{ for every } J, J' \in I_t \text{ and } \gamma \in \Gamma$$

and

$$K \in C(I_t \times I_t)_{\geq 0}^H \iff K(J, J') = K(\gamma J, \gamma J') \text{ for every } J, J' \in I_t \text{ and } \gamma \in H$$

respectively. We prove in the following that we can restrict (5.14) to kernels

$$K \in C(I_t \times I_t)^{\Gamma}_{\geq 0} \quad \text{and} \quad K' \in C(I_{t-1} \times I_{t-1})^{H}_{\geq 0}$$

by proving that for a feasible solution $(K, K')$ the vector $\left( \int_{\Gamma} K^{\gamma} d\lambda(\gamma), \int_{H} (K')^{\gamma} d\lambda(\gamma) \right)$ also yields a feasible solution with the same objective value.
First, we show that for every $\pi(P) \in I_{2t}/\Gamma \setminus \{\pi(\emptyset)\}$

$$\mathbb{1}_{I_{=1}/\Gamma}(\pi(P)) - \tilde{B}_t \left( \int_{\Gamma} K^{\gamma} d\lambda(\gamma) \right) (\pi(P)) - \tilde{B}_t^{id} \left( \int_{H} (K')^{\gamma} d\lambda(\gamma) \right) (\pi(P)) \geq 0,$$

where $h$ denotes the normalized Haar measure of the subgroup $H$. For this we observe that

$$\tilde{B}_t \left( \int_{\Gamma} K^{\gamma} d\lambda(\gamma) \right) (\pi(P)) = \int_{\Gamma} \tilde{B}_t K^{\gamma}(\pi(P)) d\lambda(\gamma)$$

and

$$\tilde{B}_t^{id} \left( \int_{H} (K')^{\gamma} d\lambda(\gamma) \right) (\pi(P)) = \int_{H} \tilde{B}_t^{id} (K')^{\gamma}(\pi(P)) dh(\gamma),$$

i.e., the Haar integrals and the operators $\tilde{B}_t$ and $\tilde{B}_t^{id}$ are interchangeable. Then we consider

$$\tilde{B}_t K^{\gamma}(\pi(P)) = \int_{\Gamma} \sum_{J, J' \in I_t: \ J \cup J' = \gamma' P} K^{\gamma}(J, J') d\lambda(\gamma')$$

$$= \int_{\Gamma} \sum_{J, J' \in I_t: \ J \cup J' = \gamma' P} K(\gamma J, \gamma J') d\lambda(\gamma')$$

$$= \int_{\Gamma} \sum_{\gamma^{-1} J, \gamma^{-1} J' \in I_t: \ \gamma^{-1}(J \cup J') = \gamma' P} K(J, J') d\lambda(\gamma')$$

$$= \int_{\Gamma} \sum_{J, J' \in I_t: \ J \cup J' = \gamma \gamma' P} K(J, J') d\lambda(\gamma')$$

$$= \tilde{B}_t K(\pi(P))$$

and conclude together with the fact that $\lambda$ is normalized that

$$\tilde{B}_t \left( \int_{\Gamma} K^{\gamma} d\lambda(\gamma) \right) (\pi(P)) = \tilde{B}_t K(\pi(P)). \tag{5.16}$$

The same arguments applied on $\tilde{B}_t^{id}(K')^\gamma(\pi(P))$ read as follows

$$\tilde{B}_t^{id}(K')^\gamma(\pi(P)) = \int_\Gamma \sum_{\substack{x\in B_r,\, J,J'\in I_{t-1}:\\ J\cup J'\cup\{x\}=\gamma'P}} K'(\gamma J, \gamma J') - \sum_{\substack{J,J'\in I_{t-1}:\\ J\cup J'=\gamma'P}} K'(\gamma J, \gamma J')\, d\lambda(\gamma')$$

$$= \int_\Gamma \sum_{\substack{\gamma^{-1}x\in B_r,\, \gamma^{-1}J,\gamma^{-1}J'\in I_{t-1}:\\ \gamma^{-1}(J\cup J'\cup\{x\})=\gamma'P}} K'(J, J') - \sum_{\substack{\gamma^{-1}J,\gamma^{-1}J'\in I_{t-1}:\\ \gamma^{-1}(J\cup J')=\gamma'P}} K'(J, J')\, d\lambda(\gamma')$$

$$= \int_\Gamma \sum_{\substack{x\in\gamma B_r,\, J,J'\in I_{t-1}:\\ J\cup J'\cup\{x\}=\gamma\gamma'P}} K'(J, J') - \sum_{\substack{J,J'\in I_{t-1}:\\ J\cup J'=\gamma\gamma'P}} K'(J, J')\, d\lambda(\gamma')$$

$$= \int_\Gamma \sum_{\substack{x\in B_r,\, J,J'\in I_{t-1}:\\ J\cup J'\cup\{x\}=\gamma\gamma'P}} K'(J, J') - \sum_{\substack{J,J'\in I_{t-1}:\\ J\cup J'=\gamma\gamma'P}} K'(J, J')\, d\lambda(\gamma')$$

$$= \tilde{B}_t^{id}K'(\pi(P)),$$

where $\gamma B_r = B_r$ holds due to the fact that $\gamma \in H$ is an isometry, and show

$$\tilde{B}_t^{id}\left(\int_H (K')^\gamma d\lambda(\gamma)\right)(\pi(P)) = \tilde{B}_t^{id}K'(\pi(P)). \tag{5.17}$$

Thus we have shown that the constraints (5.15) also hold for

$$\left(\int_\Gamma K^\gamma d\lambda(\gamma), \int_H (K')^\gamma d\lambda(\gamma)\right).$$

Since we can verify equations (5.16) and (5.17) also for $P = \emptyset$ with the above arguments, we have that the kernels $\int_\Gamma K^\gamma d\lambda(\gamma)$ and $\int_H (K')^\gamma d\lambda(\gamma)$ have the same objective value as $K$ and $K'$. This implies another reformulation of (5.14):

$$\mathcal{N}^t(X, r) = \sup \; -\tilde{B}_t K(\pi(\emptyset)) - \tilde{B}_t^{id}K'(\pi(\emptyset)) \tag{5.18}$$

$$K \in C(I_t \times I_t)_{\geq 0}^\Gamma, \; K' \in C(I_{t-1} \times I_{t-1})_{\geq 0}^H,$$

$$\mathbb{1}_{I_{=1}/\Gamma}(Q) - \tilde{B}_t K(Q) - \tilde{B}_t^{id}K'(Q) \geq 0 \text{ for all } Q \in I_{2t}/\Gamma \setminus \{\pi(\emptyset)\}.$$

## 5.7 Applications to $S^2$

In this section we aim for computing bounds for a specific covering number, namely the spherical covering number $\mathcal{N}(S^2, d)$. Here $S^2 = \{x \in \mathbb{R}^3 : x \cdot x = 1\}$ denotes the unit sphere equipped with the spherical distance $d(x, y) := \arccos(x \cdot y)$ and the unique $O(3)$-invariant measure $\omega$. As a previous lower bound we have the Coxeter-Few-Rogers bound, on which

Fejes-Tóth elaborated for $S^2$ in [31].

We fix $B_r = B(e, r)$ at the "north pole" $e = (0, 0, 1)^T$ of $S^2$ and apply hierarchy (5.18) to obtain the program

$$\mathcal{N}^t(S^2, r) = \sup \ - \tilde{B}_t K(\pi(\emptyset)) - \tilde{B}_t^{id} K'(\pi(\emptyset)) \tag{5.19}$$

$$K \in C(I_t \times I_t)_{\geq 0}^\Gamma, \ K' \in C(I_{t-1} \times I_{t-1})_{\geq 0}^H,$$

$$\mathbb{1}_{I_{=1}/\Gamma}(Q) - \tilde{B}_t K(Q) - \tilde{B}_t^{id} K'(Q) \geq 0 \text{ for all } Q \in I_{2t}/\Gamma \setminus \{\pi(\emptyset)\}.$$

Here $\Gamma = O(3)$ is the orthogonal group and $H = \text{Stab}(O(3), e)$ is the stabilizer group of $O(3)$ with respect to $e$. In the upcoming sections we formulate a finite-dimensional semidefinite program that gives lower bounds for $\mathcal{N}^2(S^2, r)$, i.e., the step $t = 2$, upto arbitrary precision. The main challenge to develop such program is to deal with the infinite-dimensional kernels $K \in C(I_2 \times I_2)_{\geq 0}^{O(3)}$ and $K' \in C(I_1 \times I_1)_{\geq 0}^{\text{Stab}(O(3), e)}$. For this purpose we will apply the famous theorem of Bochner [10] to deal with $K$ and use methods developed in an article of Bachoc and Vallentin [6] to deal with $K'$. For the general framework illustrated in the next sections we follow the lecture notes by Vallentin [80]. Further details on Bochner's theorem on compact, metrizable topological spaces $X$ equipped with a continuous action of a compact group $\Gamma$ can be found in Chapter 3 of [22].

## 5.7.1 Characterization of $K$ via Bochner's theorem

We begin this section by illustrating the theorem of Bochner for a compact metric space $X$ before we apply it to the specific case $X = S^2$. It appears convenient to have a theory, that is potentially applicable to other metric spaces as well.

The main ingredient for being able to apply Bochner's theorem is to compute a specific *complete orthonormal system* for $C(I_2)$. Let $G \subseteq \text{Aut}(I_t)$ be a subgroup of the automorphism group of $I_t$ and $\mu \in \mathcal{M}(I_2)^G$ be a $G$-invariant Radon measure on the compact metric space $(I_t, d'_H)$. A family of functions $e_1, e_2, \ldots \in C(I_2)$ is an *orthonormal system* if

$$\int_{I_2} e_k(x) e_k(x) d\mu(x) = 1, \text{ and } \int_{I_2} e_k(x) e_l(x) d\mu(x) = 0, \text{ whenever } k \neq l.$$

It is *complete* if every continuous function $f \in C(I_2)$ can be approximated arbitrarily well by finite linear combinations of $e_k$ in terms of convergence with respect to the norm

$$\|f\| := \sqrt{\int\int_{I_2} f^2(x) d\mu(x)}.$$

We aim for a complete orthonormal system that satisfies the properties given by the following theorem of Peter and Weyl (see [63] and [83]) but first, we need to recall a couple of definitions from Section 2.3:

- A subspace $S \subseteq C(X)$ is called *G-invariant* if $gS = S$ for every $g \in G$.

- A nonzero subspace $S$ is called *G-irreducible* if $\{0\}$ and $S$ are the only G-invariant subspaces of $S$.

- For two G-invariant subspaces $F, F' \subseteq C(X)$ a linear map $T : F \to F'$ is called a *G-map* if $T(gf) = \gamma T(f)$.

We further say that two spaces $S$ and $S'$ are *G-equivalent*, if there is a bijective G-map between them. Finally, we recall the theorem of Peter and Weyl.

**Theorem 5.7.1** (Peter-Weyl). *The space $C(I_t)$ has a complete orthonormal system $e_{k,i,l}$, where $k = 0, 1, \ldots$, $i = 1, 2, \ldots, m_k$, $m_k$ potentially infinite, $l = 1, 2, \ldots, h_k$, $h_k$ finite, so that*

*(1) the space $H_{k,i}$ spanned by $e_{k,i,1}, \ldots, e_{k,i,h_k}$ is G-irreducible,*

*(2) the spaces $H_{k,i}$ and $H_{k',i'}$ are G-equivalent if and only if $k = k'$,*

*(3) there are G-maps $\phi_{k,i} : H_{k,1} \to H_{k,i}$ mapping $e_{k,1,l}$ to $e_{k,i,l}$.*

The proof of this theorem is constructive in the sense that one constructs an orthonormal system satisfying (1) – (3), thus we can apply Bochner's Theorem [10].

**Theorem 5.7.2** (Bochner). *Let $e_{k,i,l}$ be a complete orthonormal system for $C(I_t)$ as in Theorem 5.7.1. Every G-invariant, positive semidefinite kernel $K \in C(I_t \times I_t)_{\geq 0}^{G}$ can be written as*

$$K(J, J') = \sum_{k=0}^{\infty} \sum_{i,j=1}^{m_k} f_{k,ij} \sum_{l=1}^{h_k} e_{k,i,l}(J) e_{k,j,l}(J'), \qquad (5.20)$$

*or more economically as*

$$K(J, J') = \sum_{k=0}^{\infty} \langle F_k, Z_k^{(J,J')} \rangle, \qquad (5.21)$$

*with $(F_k)_{ij} = f_{k,ij}$ and $\left( Z_k^{(J,J')} \right)_{ij} = \sum_{l=1}^{h_k} e_{k,i,l}(J) e_{k,j,l}(J')$. Here, $F_k$ is symmetric and positive semidefinite. If G is transitive, the series $\sum_{k=0}^{\infty} \langle F_k, Z_k^{(J,J')} \rangle$ converges absolutely-uniformly.*

De Laat [22] generalized this theory slightly and proved the convergence of the series $\sum_{k=0}^{\infty} \langle F_k, Z_k^{(J,J')} \rangle$ for G having finitely many orbits. His characterization of the kernels $Z_k^{(J,J')}$ uses representation theory, in particular *unitary representations of G*, which are continuous group homomorphisms $\rho : G \to U(\mathcal{H})$ valued on the group of unitary operators on a nonzero Hilbert space $\mathcal{H}$, given that $U(\mathcal{H})$ is endowed with the strong operator topology. He concluded that for a complete set $\widehat{G}$ of non-equivalent, irreducible unitary representations of G, we have a block $Z_\rho$ for every representation $\rho \in \widehat{G}$ called a *zonal matrix*. The obstacle that the zonal matrices $Z_\rho \in S_{\geq 0}^{m_\rho}$ are possibly infinite-dimensional can be dealt with by considering $R_{\rho,d} \times R_{\rho,d}$ submatrices, where $R_{\rho,1} \subseteq R_{\rho,2} \subseteq \ldots [m_\rho]$ are finite subsets such that $\bigcup_{d=0}^{\infty} R_{\rho,d} = [m_\rho]$ and for every $d$ the set $R_{\rho,d}$ is empty for all but finitely many $\rho$.

From now on, we will focus on the specific space $X = S^2$. By computing concrete zonal submatrices in Section 7.6.3 of [22], one achieves an inner approximation of the cone $C(I_2 \times I_2)_{\geq 0}^{O(3)}$ that was used to approach the *energy minimization problem*. Here, the sequence

$$C_1 \subseteq C_2 \subseteq \ldots \subseteq C(I_2 \times I_2)_{\geq 0}^{O(3)}$$

defined by

$$C_d = \left\{ \sum_{\rho \in \widehat{O(3)}} \langle F_\rho, Z_{\rho,d}(.,.) \rangle \; : \; F_\rho \in S_{\geq 0}^{R_{\rho,d}} \text{ for } \rho \in \widehat{O(3)} \right\}$$

approximates the cone $C(I_2 \times I_2)_{\geq 0}^{O(3)}$ and it can be shown that $\bigcup_{d=1}^{\infty} C_d$ is uniformly dense in $C(I_2 \times I_2)_{\geq 0}^{O(3)}$. Ultimately, this led to the first computation of the second step of the Lasserre hierarchy for an infinite-dimensional problem.

On the other hand, the problem formulation of energy minimization does only rely on one kernel $K \in C(I_2 \times I_2)_{\geq 0}^{O(3)}$. Although de Laat's theory would also apply on $C(I_1 \times I_1)_{\geq 0}^{\text{Stab}(O(3),e)}$ we use the paper of [6] to find an inner approximation scheme for $K'$, since they provide concrete approximating functions for $K'$.

## 5.7.2   Explicit description of $K'$ via Bachoc-Vallentin's theorem

The approximation scheme for $K'$ relies on a result of Bachoc and Vallentin [6] on the cone $C(S^n \times S^n)_{\geq 0}^{\text{Stab}(O(n+1),e)}$.

Suppose that the space $C(I_1)$ is equipped with a $\text{Stab}(O(3), e)$-invariant Radon measure $\mu \in M(I_1)^{\text{Stab}(O(3),e)}$ scaled to $\mu(\{\emptyset\}) = 1$ (which exists if (5.9) is non-empty). Then, we first observe that $C(I_1)$ can be split into two parts

$$C(I_1) = C(I_{=0}) \perp C(I_{=1}) \cong \mathbb{R} \perp C(S^2).$$

Next, one constructs a complete orthonormal system for $C(I_{=0}) \perp C(I_{=1})$. For $C(I_{=0})$ we set

$$e_{0,-1,1}(J) = \begin{cases} 1 & \text{if } J = \emptyset \\ 0 & \text{if } J \in I_{=1} \end{cases}$$

and construct $e_{k,i,l}$ for $C(S^2)$ by following Bachoc and Vallentin (see Theorem 6.8 in [6]). Then we identify $e_{k,i,l}(\{x\}) = e_{k,i,l}(x)$ for $x \in S^2$ and extend the orthonormal system with $e_{k,i,l}(\emptyset) = 0$ to keep orthogonality with $e_{0,-1,1}$.

By applying Bochner's Theorem, we achieve an approximate formulation for the kernel $K' \in C(I_1 \times I_1)_{\geq 0}^{\text{Stab}(O(3),e)}$ by defining

$$K_d'(J, J') := \sum_{i,j=-1}^{d} f_{0,ij} Y_{0,ij}(J, J') + \sum_{k=1}^{d} \sum_{i,j=0}^{d-k} f_{k,ij} Y_{k,ij}(J, J'), \tag{5.22}$$

where

$$Y_0(J, J') := \begin{pmatrix} e_{0,-1,1}(J) \\ \vdots \\ e_{0,d,1}(J) \end{pmatrix} \begin{pmatrix} e_{0,-1,1}(J') \\ \vdots \\ e_{0,d,1}(J') \end{pmatrix}^T ,$$

with the exact values of $e_{0,i,1}$ that can be found in [3] Chapter 9.8, see also the proof of Theorem 3.2 in [6]. For $u = u(\{x\}, \{y\}) = x \cdot e$, $v = v(\{x\}, \{y\}) = y \cdot e$ and $w = w(\{x\}, \{y\}) = x \cdot y$, we define the corresponding matrices $Y_k(J, J')$ with $k = 1, \dots, d$ explicitly by

$$Y_{k,ij}(J, J') = \begin{cases} u^i v^j Q_k^2(u, v, w) & \text{if } J, J' \in I_{=1}, \\ 0 & \text{otherwise.} \end{cases}$$

The $Q_k^2$ are defined via the *Gegenbauer polynomials* $C_k^0$ of degree $k$:

$$Q_k^2(u, v, w) = \left( (1 - u^2)(1 - v^2) \right)^{\frac{k}{2}} C_k^0 \left( \frac{w - uv}{\sqrt{(1 - u^2)(1 - v^2)}} \right) \Big/ C_k^0(1).$$

Gegenbauer or ultraspherical polynomials $C_k^\lambda$ occur as orthogonal polynomials in a variety of situations and are well established objects. For more details we refer to [3]. The sequence of kernels $K_d'$ defines a sequence of cones

$$C_d' = \left\{ \langle F_0, Y_0(.,.)\rangle + \sum_{k=1}^{d} \langle F_k, Y_k(.,.)\rangle : F_0 \in S_{\geq 0}^{d+2}, F_k \in S_{\geq 0}^{d-k+1} \text{ for } k \in \{1, \dots, d\} \right\}$$

having the property

$$C_1' \subseteq C_2' \subseteq \dots \subseteq C(I_1 \times I_1)_{\geq 0}^{\text{Stab}(O(3),e)}.$$

Finally, we conclude that $\bigcup_{d=0}^{\infty} C_d'$ is dense in $C(I_1 \times I_1)_{\geq 0}^{\text{Stab}(O(3),e)}$ by the Stone-Weierstrass theorem. The semidefinite program that provides a lower bound for $\mathcal{N}^2(S^2, r)$ can finally be stated as

$$\mathcal{N}^2(S^2, r) \geq \sup \ - \tilde{B}_2 K(\pi(\emptyset)) - \tilde{B}_2^{id} K'(\pi(\emptyset)) \tag{5.23}$$
$$K \in C_d, \ K' \in C_d',$$
$$\mathbb{1}_{I_{=1}/\Gamma}(Q) - \tilde{B}_2 K(Q) - \tilde{B}_2^{id} K'(Q) \geq 0 \text{ for all } Q \in I_{2t}/\Gamma \setminus \{\pi(\emptyset)\}.$$

and provides increasingly better bounds for growing values of $d$. The drawback of this method is that increasing $d$ also increases the dimension of the underlying matrices and hence the computation time. Thus it remains to determine an appropriate value for $d$, which is a challenging task for future research.

Also, one observes that if one simply takes a finite sample of constraints $Q \in I_{2t}/\Gamma \setminus \{\pi(\emptyset)\}$, this does not necessarily yield a feasible solution for the original problem. However, since the functions $\mathbb{1}_{l_{=1}/\Gamma,}$, $\tilde{B}_t K$ and $\tilde{B}_t^{id} K'$ are continuous and bounded, a feasible solution $(K, K')$ to our sampled problem is almost feasible, i.e., there is a small constant $c > 0$ such that

$$\mathbb{1}_{l_{=1}/\Gamma}(Q) - \tilde{B}_2 K(Q) - \tilde{B}_2^{id} K'(Q) \geq -c \text{ for all } Q \in I_{2t}/\Gamma \setminus \{\pi(\emptyset)\}.$$

In related problems such as *packing* (see, e.g., [28]), it is often possible that one can determine additional kernels $K^+ \in C(I_2 \times I_2)_{\geq 0}^{\bar{O}(3)}$ or $(K')^+ \in C(I_1 \times I_1)_{\geq 0}^{\text{Stab}(O(3),e)}$ such that $(K + K^+, K' + (K')^+)$ is a feasible solution for the original problem $\mathcal{N}^2(S^2, r)$. Such kernels are still to be found for the covering number – another task for possible future research on this very interesting topic.

# Appendix

In this appendix we cite the actual values for the covering radius $r$ (Column 2) corresponding to a fixed number of caps (Column 1) on Neil Sloane's webpage [76]. Additionally, Column 3 shows the lower bounds given by L. Fejes-Tóth [31] for the covering number $\mathcal{N}(S^2, r)$ with covering radius $r$.

| number of caps | covering radius $r$ | Fejes-Tóth Bound |
| :---: | :---: | :---: |
| 4 | 70.5287794 | 3, 999999997 |
| 5 | 63.4349488 | 4.697992631 |
| 6 | 54.7356103 | 6, 000000003 |
| 7 | 51.0265527 | 6, 780802809 |
| 8 | 48.1395291 | 7, 521978584 |
| 9 | 45.8788878 | 8, 205126707 |
| 10 | 42.3078266 | 9.519950643 |
| 11 | 41.4271960 | 9.898287382 |
| 12 | 37.3773681 | 12.00000002 |
| 13 | 37.0685427 | 12.18933749 |
| 14 | 34.9379270 | 13.63625733 |
| 15 | 34.0399001 | 14.32949147 |
| 16 | 32.8988128 | 15.29393600 |
| 17 | 32.0929328 | 16.03817023 |
| 18 | 31.0131718 | 17.12803589 |
| 19 | 30.3686748 | 17.83484936 |
| 20 | 29.6230958 | 18.71091588 |

Table 5.1: Values from [76] and [31]

| number of caps | covering radius $r$ | Fejes-Tóth Bound |
|---|---|---|
| 21 | 28.8244768 | 19.72591251 |
| 22 | 27.8100588 | 21.14365366 |
| 23 | 27.4818687 | 21.63634148 |
| 24 | 26.8126364 | 22.69774615 |
| 25 | 26.3287855 | 23.51614618 |
| 26 | 25.8449223 | 24.38102783 |
| 27 | 25.2509549 | 25.51152965 |
| 28 | 24.6589489 | 26.72065715 |
| 29 | 24.3683986 | 27.34661483 |
| 30 | 23.8787580 | 28.45369863 |
| 31 | 23.6119921 | 29.08607870 |
| 32 | 22.6904804 | 31.44496901 |
| 33 | 22.5905116 | 31.71838131 |
| 34 | 22.3314637 | 32.44404754 |
| 35 | 22.0725569 | 33.19501458 |
| 36 | 21.6994390 | 34.32493267 |
| 37 | 21.3100299 | 35.56808878 |
| 38 | 21.0698584 | 36.36943712 |
| 39 | 20.8511244 | 37.12349474 |
| 40 | 20.4721353 | 38.48771501 |
| 41 | 20.3177152 | 39.06559337 |
| 42 | 20.0480917 | 40.10680322 |
| 43 | 19.8428333 | 40.92808274 |
| 44 | 19.6375705 | 41.77527880 |
| 45 | 19.4207407 | 42.69956370 |
| 46 | 19.1586113 | 43.85913546 |
| 47 | 18.9924594 | 44.61913513 |
| 48 | 18.6892566 | 46.05864541 |
| 49 | 18.5926796 | 46.53202118 |
| 50 | 18.3000226 | 48.01255497 |

Table 5.2: Values from [76] and [31]

| number of caps | covering radius $r$ | Fejes-Tóth Bound |
|---|---|---|
| 51 | 18.1990011 | 48.54027920 |
| 52 | 18.0544758 | 49.31072403 |
| 53 | 17.8845734 | 50.24045050 |
| 54 | 17.6791447 | 51.40058493 |
| 55 | 17.5222392 | 52.31431798 |
| 56 | 17.3501139 | 53.34534073 |
| 57 | 17.1758476 | 54.42093237 |
| 58 | 17.0199610 | 55.41121754 |
| 59 | 16.9034031 | 56.16963721 |
| 60 | 16.7719330 | 57.04413921 |
| 61 | 16.6391845 | 57.94826496 |
| 62 | 16.4906596 | 58.98583938 |
| 63 | 16.3679364 | 59.86456754 |
| 64 | 16.1940190 | 61.14424743 |
| 65 | 16.1114061 | 61.76668239 |
| 66 | 15.9550615 | 62.97121529 |
| 67 | 15.8581808 | 63.73556439 |
| 68 | 15.7236959 | 64.82010860 |
| 69 | 15.5950401 | 65.88401852 |
| 70 | 15.4951288 | 66.72857619 |
| 71 | 15.3918904 | 67.61859506 |
| 72 | 15.1445321 | 69.82562079 |
| 73 | 15.1164437 | 70.08310996 |
| 74 | 15.0311866 | 70.87353121 |
| 75 | 14.9454277 | 71.68229221 |
| 76 | 14.8539208 | 72.56075697 |
| 77 | 14.7449905 | 73.62788608 |
| 78 | 14.6550577 | 74.52690100 |
| 79 | 14.5627673 | 75.46685649 |
| 80 | 14.4503042 | 76.63670271 |

Table 5.3: Values from [76] and [31]

| number of caps | covering radius $r$ | Fejes-Tóth Bound |
|:---:|:---:|:---:|
| 81 | 14.3767803 | 77.41638875 |
| 82 | 14.2863117 | 78.39233514 |
| 83 | 14.2239571 | 79.07586582 |
| 84 | 14.1157901 | 80.28315034 |
| 85 | 14.0452618 | 81.08540675 |
| 86 | 13.9626271 | 82.04088444 |
| 87 | 13.8849703 | 82.95439891 |
| 88 | 13.7904978 | 84.08660049 |
| 89 | 13.7120948 | 85.04404307 |
| 90 | 13.6208737 | 86.17889502 |
| 91 | 13.5633748 | 86.90601601 |
| 92 | 13.4878634 | 87.87508267 |
| 93 | 13.4258226 | 88.68354322 |
| 94 | 13.3486301 | 89.70522923 |
| 95 | 13.2858097 | 90.54987052 |
| 96 | 13.2112886 | 91.56749885 |
| 97 | 13.1421390 | 92.52730359 |
| 98 | 13.0644481 | 93.62389667 |
| 99 | 12.9972791 | 94.58786934 |
| 100 | 12.9360973 | 95.47901769 |
| 101 | 12.8693268 | 96.46611057 |
| 102 | 12.8065480 | 97.40831141 |
| 103 | 12.7396985 | 98.42695978 |
| 104 | 12.6710007 | 99.49061599 |
| 105 | 12.6206479 | 100.2812910 |
| 106 | 12.5580705 | 101.2772116 |
| 107 | 12.4984676 | 102.2397371 |
| 108 | 12.4268411 | 103.4148046 |
| 109 | 12.3823879 | 104.1543586 |
| 110 | 12.3000527 | 105.5453845 |

Table 5.4: Values from [76] and [31]

| number of caps | covering radius $r$ | Fejes-Tóth Bound |
|:---:|:---:|:---:|
| 111 | 12.2463574 | 106.4677036 |
| 112 | 12.1906904 | 107.4367870 |
| 113 | 12.1475147 | 108.1976069 |
| 114 | 12.0965651 | 109.1059149 |
| 115 | 12.0509312 | 109.9292569 |
| 116 | 11.9886435 | 111.0682838 |
| 117 | 11.9433897 | 111.9070210 |
| 118 | 11.8858175 | 112.9879469 |
| 119 | 11.8437244 | 113.7882467 |
| 120 | 11.7866486 | 114.8871328 |
| 121 | 11.7339186 | 115.9166330 |
| 122 | 11.6770714 | 117.0421779 |
| 123 | 11.6364348 | 117.8568923 |
| 124 | 11.5887833 | 118.8231863 |
| 125 | 11.5384928 | 119.8560078 |
| 126 | 11.4894028 | 120.8772839 |
| 127 | 11.4535231 | 121.6320522 |
| 128 | 11.4068506 | 122.6245375 |
| 129 | 11.3563380 | 123.7125009 |
| 130 | 11.3165625 | 124.5794765 |

Table 5.5: Values from [76] and [31]

# Bibliography

[1] T. Acisu. *Approximating Set Cover using the Lasserre hierarchy*. Bachelor's thesis. University of Cologne, 2018.

[2] L. Ambrosio, N. Gigli, and G. Savaré. *Gradient flows in metric spaces and in the space of probability measures*. Lectures in Mathematics ETH Zürich. Birkhäuser Verlag, Basel, 2005.

[3] G. E. Andrews, R. Askey, and R. Roy. *Special functions*, volume 71 of *Encyclopedia of Mathematics and its Applications*. Cambridge University Press, Cambridge, 1999.

[4] S. Artstein-Avidan and O. Raz. Weighted covering numbers of convex sets. *Adv. Math.*, 227(1):730–744, 2011.

[5] S. Artstein-Avidan and B. A. Slomka. On weighted covering numbers and the Levi-Hadwiger conjecture. *Israel J. Math.*, 209(1):125–155, 2015.

[6] C. Bachoc and F. Vallentin. New upper bounds for kissing numbers from semidefinite programming. *J. Amer. Math. Soc.*, 21(3):909–924, 2008.

[7] R. P. Bambah. On lattice coverings by spheres. In *Proc. Nat. Inst. Sci. India*, volume 20, pages 25–52, 1954.

[8] A. Barvinok. *A course in convexity*, volume 54 of *Graduate Studies in Mathematics*. American Mathematical Society, Providence, RI, 2002.

[9] H. Bauer. *Maß- und Integrationstheorie*. de Gruyter Lehrbuch. [de Gruyter Textbook]. Walter de Gruyter & Co., Berlin, second edition, 1992.

[10] S. Bochner. Hilbert distances and positive definite functions. *Ann. of Math. (2)*, 42:647–656, 1941.

[11] K. Böröczky, Jr. and G. Wintsche. Covering the sphere by equal spherical balls. In *Discrete and computational geometry*, volume 25 of *Algorithms Combin.*, pages 235–251. Springer, Berlin, 2003.

[12] E. Chlamtáč, Z. Friggstad, and K. Georgiou. Lift-and-project methods for set cover and knapsack. In *Algorithms and data structures*, volume 8037 of *Lecture Notes in Comput. Sci.*, pages 256–267. Springer, Heidelberg, 2013.

[13] V. Chvátal. A greedy heuristic for the set-covering problem. *Math. Oper. Res.*, 4(3):233–235, 1979.

[14] Jaka Cimprič, Salma Kuhlmann, and Claus Scheiderer. Sums of squares and moment problems in equivariant situations. *Trans. Amer. Math. Soc.*, 361(2):735–765, 2009.

[15] J. H. Conway and N. J. A. Sloane. *Sphere packings, lattices and groups*, volume 290 of *Grundlehren der Mathematischen Wissenschaften [Fundamental Principles of Mathematical Sciences]*. Springer-Verlag, New York, third edition, 1999. With additional contributions by E. Bannai, R. E. Borcherds, J. Leech, S. P. Norton, A. M. Odlyzko, R. A. Parker, L. Queen and B. B. Venkov.

[16] H. S. M. Coxeter, L. Few, and C. A. Rogers. Covering space with equal spheres. *Mathematika*, 6:147–157, 1959.

[17] F. Cucker and S. Smale. On the mathematical foundations of learning. *Bull. Amer. Math. Soc. (N.S.)*, 39(1):1–49, 2002.

[18] G. B. Dantzig. Maximization of a linear function of variables subject to linear inequalities. In *Activity Analysis of Production and Allocation*, Cowles Commission Monograph No. 13, pages 339–347. John Wiley & Sons, Inc., New York, N. Y.; Chapman & Hall, Ltd., London, 1951.

[19] G. B. Dantzig. *Linear programming and extensions*. Princeton Landmarks in Mathematics. Princeton University Press, Princeton, NJ, corrected edition, 1998.

[20] E. de Klerk and F. Vallentin. On the Turing model complexity of interior point methods for semidefinite programming. *SIAM J. Optim.*, 26(3):1944–1961, 2016.

[21] D. de Laat. Moment methods in energy minimization: New bounds for riesz minimal energy problems. 2016.

[22] D. de Laat. *Moment methods in extremal geometry*. PHD thesis. Technical University of Delft, 2016.

[23] D. de Laat and F. Vallentin. A semidefinite programming hierarchy for packing problems in discrete geometry. *Math. Program.*, 151(2, Ser. B):529–553, 2015.

[24] A. Deitmar and S. Echterhoff. *Principles of harmonic analysis*. Universitext. Springer, New York, 2009.

[25] B. N. Delone and S. S. Ryškov. Solution of the problem on the least dense lattice covering of a 4-dimensional space by equal spheres. *Dokl. Akad. Nauk SSSR*, 152:523–524, 1963.

[26] J. Dieudonné. Sur la séparation des ensembles convexes. *Math. Ann.*, 163:1–3, 1966.

[27] I. Dinur and D. Steurer. Analytical approach to parallel repetition. In *STOC'14—Proceedings of the 2014 ACM Symposium on Theory of Computing*, pages 624–633. ACM, New York, 2014.

[28] M. Dostert, C. Guzmán, F. M. de Oliveira Filho, and F. Vallentin. New upper bounds for the density of translative packings of three-dimensional convex bodies with tetrahedral symmetry. *Discrete Comput. Geom.*, 58(2):449–481, 2017.

[29] I. Dumer. Covering spheres with spheres. *Discrete Comput. Geom.*, 38(4):665–679, 2007.

[30] G. Fejes Tóth. A note on covering by convex bodies. *Canad. Math. Bull.*, 52(3):361–365, 2009.

[31] L. Fejes-Tóth. *Lagerungen in der Ebene auf der Kugel und im Raum*. Grundlehren der mathematischen Wissenschaften in Einzeldarstellungen mit besonderer Berücksichtigung der Anwendungsgebiete. Springer, 1953.

[32] G. B. Folland. *A course in abstract harmonic analysis*. Studies in Advanced Mathematics. CRC Press, Boca Raton, FL, 1995.

[33] G. B. Folland. *Real analysis*. Pure and Applied Mathematics (New York). John Wiley & Sons, Inc., New York, second edition, 1999. Modern techniques and their applications, A Wiley-Interscience Publication.

[34] S. Foucart and H. Rauhut. *A mathematical introduction to compressive sensing*. Applied and Numerical Harmonic Analysis. Birkhäuser/Springer, New York, 2013.

[35] D. C. Gijswijt, H. D. Mittelmann, and A. Schrijver. Semidefinite code bounds based on quadruple distances. *IEEE Trans. Inform. Theory*, 58(5):2697–2705, 2012.

[36] D.C. Gijswijt. Matrix algebras and semidefinite programming techniques for codes. 2005.

[37] M. X. Goemans and D. P. Williamson. Improved approximation algorithms for maximum cut and satisfiability problems using semidefinite programming. *J. Assoc. Comput. Mach.*, 42(6):1115–1145, 1995.

[38] H. Groemer. Existenzsätze für Lagerungen im Euklidischen Raum. *Mathematische Zeitschrift*, 81:260–278, 1963.

[39] M. Grötschel, L. Lovász, and A. Schrijver. The ellipsoid method and its consequences in combinatorial optimization. *Combinatorica*, 1(2):169–197, 1981.

[40] M. Grötschel, L. Lovász, and A. Schrijver. Polynomial algorithms for perfect graphs. In *Topics on perfect graphs*, volume 88 of *North-Holland Math. Stud.*, pages 325–356. North-Holland, Amsterdam, 1984.

[41] M. Grötschel, L. Lovász, and A. Schrijver. *Geometric algorithms and combinatorial optimization*, volume 2 of *Algorithms and Combinatorics: Study and Research Texts*. Springer-Verlag, Berlin, 1988.

[42] P. M. Gruber. *Convex and discrete geometry*, volume 336 of *Grundlehren der Mathematischen Wissenschaften [Fundamental Principles of Mathematical Sciences]*. Springer, Berlin, 2007.

[43] D. Handel. Some homotopy properties of spaces of finite subsets of topological spaces. *Houston J. Math.*, 26(4):747–764, 2000.

[44] A. Hatcher. *Algebraic topology*. Cambridge University Press, Cambridge, 2002.

[45] T. W. Haynes, S. T. Hedetniemi, and P. J. Slater. *Fundamentals of domination in graphs*, volume 208 of *Monographs and Textbooks in Pure and Applied Mathematics*. Marcel Dekker, Inc., New York, 1998.

[46] K. Jänich. *Topologie*. Springer-Lehrbuch. [Springer Textbook]. Springer-Verlag, Berlin, fourth edition, 1994. With a chapter by Theodor Bröcker.

[47] D. S. Johnson. Approximation algorithms for combinatorial problems. *J. Comput. System Sci.*, 9:256–278, 1974. Fifth Annual ACM Symposium on the Theory of Computing (Austin, Tex., 1973).

[48] L. V. Kantorovich. Mathematical methods of organizing and planning production. *Management Sci.*, 6:366–422, 1959/1960.

[49] R. M. Karp. Reducibility among combinatorial problems. pages 85–103, 1972.

[50] J. L. Kelley. *General topology*. D. Van Nostrand Company, Inc., Toronto-New York-London, 1955.

[51] R. Kershner. The number of circles covering a set. *American Journal of Mathematics*, 61(3):665–671, 1939.

[52] V. L. Klee, Jr. Separation properties of convex cones. *Proc. Amer. Math. Soc.*, 6:313–318, 1955.

[53] J. B. Lasserre. Global optimization with polynomials and the problem of moments. *SIAM J. Optim.*, 11(3):796–817, 2000/01.

[54] J. B. Lasserre. An explicit equivalent positive semidefinite program for nonlinear 0-1 programs. *SIAM J. Optim.*, 12(3):756–769, 2002.

[55] M. Ledoux and M. Talagrand. *Probability in Banach spaces*, volume 23 of *Ergebnisse der Mathematik und ihrer Grenzgebiete (3) [Results in Mathematics and Related Areas (3)]*. Springer-Verlag, Berlin, 1991. Isoperimetry and processes.

[56] F. W. Levi. Überdeckung eines Eibereiches durch Parallelverschiebung seines offenen Kerns. *Arch. Math. (Basel)*, 6:369–370, 1955.

[57] L. Lovász. On the ratio of optimal integral and fractional covers. *Discrete Math.*, 13(4):383–390, 1975.

[58] P. Mattila. *Geometry of sets and measures in Euclidean spaces*, volume 44 of *Cambridge Studies in Advanced Mathematics*. Cambridge University Press, Cambridge, 1995. Fractals and rectifiability.

[59] M. Naszódi. Flavors of translative coverings. 2016.

[60] M. Naszódi. On some covering problems in geometry. *Proc. Amer. Math. Soc.*, 144(8):3555–3562, 2016.

[61] M. A. Nielsen and I. L. Chuang. *Quantum computation and quantum information*. Cambridge University Press, Cambridge, 2000.

[62] R. O'Donnell. SOS is not obviously automatizable, even approximately. In *8th Innovations in Theoretical Computer Science Conference*, volume 67 of *LIPIcs. Leibniz Int. Proc. Inform.*, pages Art. No. 59, 10. Schloss Dagstuhl. Leibniz-Zent. Inform., Wadern, 2017.

[63] F. Peter and H. Weyl. Die Vollständigkeit der primitiven Darstellungen einer geschlossenen kontinuierlichen Gruppe. *Math. Ann.*, 97(1):737–755, 1927.

[64] A. Reznikov and E. B. Saff. The covering radius of randomly distributed points on a manifold. *Int. Math. Res. Not. IMRN*, (19):6065–6094, 2016.

[65] C. Riener, T. Theobald, L. Jansson Andrén, and J. B. Lasserre. Exploiting symmetries in SDP-relaxations for polynomial optimization. *Math. Oper. Res.*, 38(1):122–141, 2013.

[66] C. A. Rogers. A note on coverings. *Mathematika*, 4:1–6, 1957.

[67] C. A. Rogers. Covering a sphere with spheres. *Mathematika*, 10:157–164, 1963.

[68] C. A. Rogers. *Packing and covering*. Cambridge Tracts in Mathematics and Mathematical Physics, No. 54. Cambridge University Press, New York, 1964.

[69] J. H. Rolfes. *Copositive Formulations of the Dominating Set problem and applications*. Master's thesis. University of Cologne, 2014.

[70] J. H. Rolfes and F. Vallentin. Covering compact metric spaces greedily. *Acta Math. Hungar.*, 155(1):130–140, 2018.

[71] T. Rothvoß. Lecture notes: The lasserre hierarchy in approximation algorithms. Technical report, 2013.

[72] S. S. Ryškov and E. P. Baranovskiĭ. Solution of the problem of the least dense lattice covering of five-dimensional space by equal spheres. *Dokl. Akad. Nauk SSSR*, 222(1):39–42, 1975.

[73] I. J. Schoenberg. Positive definite functions on spheres. *Duke Math. J.*, 9:96–108, 1942.

[74] A. Schrijver. *Theory of linear and integer programming*. Wiley-Interscience Series in Discrete Mathematics. John Wiley & Sons, Ltd., Chichester, 1986. A Wiley-Interscience Publication.

[75] B. Simon. *Convexity*, volume 187 of *Cambridge Tracts in Mathematics*. Cambridge University Press, Cambridge, 2011. An analytic viewpoint.

[76] N. J. A. Sloane. Spherical coverings. Technical report, 2015.

[77] E. M. Stein and R. Shakarchi. *Functional analysis*, volume 4 of *Princeton Lectures in Analysis*. Princeton University Press, Princeton, NJ, 2011. Introduction to further topics in analysis.

[78] S. K. Stein. Two combinatorial covering theorems. *J. Combinatorial Theory Ser. A*, 16:391–397, 1974.

[79] A. Thue. Om nogle geometrisk-taltheoretiske theoremer. *Forhandlingerne ved de Skandinaviske Naturforskeres*, 14:352–353, 1892.

[80] F. Vallentin. Lecture notes: Semidefinite programs and harmonic analysis. Technical report, 2008.

[81] J. von Neumann. Discussion of a maximum problem. In *Collected works. Vol. VI: Theory of games, astrophysics, hydrodynamics and meteorology*, General editor: A. H. Taub. A Pergamon Press Book, pages 89–95. The Macmillan Co., New York, 1963.

[82] D. Werner. *Funktionalanalysis*. Springer-Lehrbuch. Springer Berlin Heidelberg, 2007.

[83] H. Weyl. Harmonics on homogeneous manifolds. *Ann. of Math. (2)*, 35(3):486–499, 1934.

# Index

# Publication list

**In preparation**

- *An SDP hierarchy for the covering problem.* David de Laat, Cordian Riener, Jan Hendrik Rolfes, Frank Vallentin

**Published**

- *Covering compact metric spaces greedily.* Jan Hendrik Rolfes, Frank Vallentin, 11 pages, published in Acta Mathematica Hungarica, May 2018, https://link.springer.com/article/10.1007/s10474-018-0829-4

**Surveys**

- *Das Problem der Kugelpackung (in German).* Maria Dostert, Stefan Krupp, Jan Rolfes, 12 pages, published in Snapshot of modern mathematics from Oberwolfach, April 2016, https://imaginary.org/snapshot/das-problem-der-kugelpackung

# Erklärung

Ich versichere, dass ich die von mir vorgelegte Dissertation selbständig angefertigt, die benutzten Quellen und Hilfsmittel vollständig angegeben und die Stellen der Arbeit - einschließlich Tabellen, Karten und Abbildungen -, die anderen Werken im Wortlaut oder dem Sinn nach entnommen sind, in jedem Einzelfall als Entlehnung kenntlich gemacht habe; dass diese Dissertation noch keiner anderen Fakultät oder Universität zur Prüfung vorgelegen hat; dass sie - abgesehen von unten angegebenen Teilpublikationen - noch nicht veröffentlicht worden ist sowie, dass ich eine solche Veröffentlichung vor Abschluss des Promotionsverfahrens nicht vornehmen werde. Die Bestimmungen der Promotionsordnung sind mir bekannt. Die von mir vorgelegte Dissertation ist von Prof. Dr. Frank Vallentin betreut worden.

---

Jan Hendrik Rolfes

**Teilpublikationen:**

- *Das Problem der Kugelpackung (in German).* Maria Dostert, Stefan Krupp, Jan Rolfes, Snapshot of modern mathematics from Oberwolfach, April 2016, 12 pages, https://imaginary.org/snapshot/das-problem-der-kugelpackung

- *Covering compact metric spaces greedily.* Jan Hendrik Rolfes, Frank Vallentin, Acta Mathematica Hungarica, May 2018, 11 pages, https://link.springer.com/article/10.1007/s10474-018-0829-4

# Lebenslauf

**Persönliche Daten:**

|  |  |
|---|---|
| Name | Jan Hendrik Rolfes |
| Geboren | 06.08.1988 in Neuwied |
| Staatsangehörigkeit | deutsch |

**Studienverlauf:**

Seit 02/2015    Promotionsstudium an der Universität zu Köln
Betreuer: Prof. Dr. Frank Vallentin

04/2012 – 12/2014    Studium an der Universität zu Köln
Fachrichtung: Wirtschaftsmathematik
Abschluss: Master of Science

10/2008 – 04/2012    Studium an der Universität zu Köln
Fachrichtung: Wirtschaftsmathematik
Abschluss: Bachelor of Science

**Zivildienst:**

04/2008 – 10/2008    Heinrich-Haus Gruppe Engers

**Schulbildung:**

08/1999 – 03/2008    Werner Heisenberg Gymnasium Neuwied
Abschluss: Abitur

**Beruflicher Werdegang:**

02/2017 – 01/2019    Wissenschaftliche Hilfskraft
Universität zu Köln

02/2016 – 01/2017    Wissenschaftlicher Mitarbeiter
Universität zu Köln

02/2015 – 01/2016    Wissenschaftliche Hilfskraft
Universität zu Köln